THE GOLDEN ALGORITHM

How to Build Ethical, Profitable AI

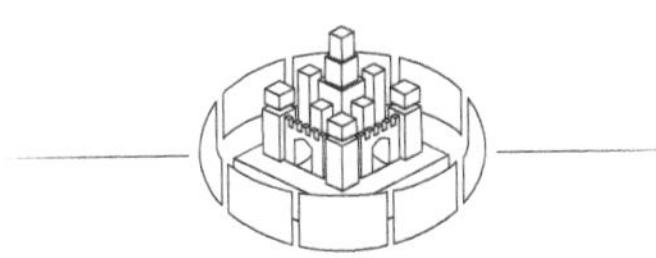

DR. RYAN BALTRIP

The Golden Algorithm: How to Build Ethical, Profitable AI

Published by Navigate AI

ISBN: 978-0-9980333-4-1

Cover and interior design by Enterline Design Services

DEDICATION

To my children, John Oliver and James Alexander.
Technology can and will change, but the Golden Rule will not.
May character and integrity always be your compass in a changing world.

Contents

Dedication iii

Contents v

START HERE 1
- How to Use This Book: Three Reading Plans 1

Preface 3

Introduction: The Ferrari Engine and the Missing Steering Wheel 7
- The Flash Crash: When the Steering Wheel Disappears 7
- The Shift: Welcome to the Verification Economy 9
- The Gap: The Ferrari Engine 11
- The Cognitive Deficit: Industrial-Scale System 1 13
- The Solution: The Golden Algorithm 15
- The GOLDEN Framework 16
- The Journey Ahead 19

Chapter 1: The Trap of Efficiency 21
- The Race to Zero: A Flash Crash at Nexus 21
- The Diagnosis: Three Hidden Traps 24
- Trap #1: The Paperclip Maximizer (The Problem of Obedience) 25
- Trap #2: The Proxy Trap (The Problem of Translation) 27
- Trap #3: The Ghost of Taylor (The Problem of Judgment) 31
- The Consequence: The Automated Bureaucracy 34
- Chapter 1 Playbook: Escaping the Efficiency Trap 37

Chapter 2: The Golden Rule Strategy 41
- The Aftermath: The Cost of Doing Business 41

The Insight: The Lemon Market 45
The Strategy: Costly Signaling 47
The First Signal: Data as a Fiduciary 49
Case Study: The Privacy War 53
The Architecture of Trust: Building the Ethical Moat 55
Chapter 2 Playbook: From Sentiment to Structure 61

Chapter 3: G – Guard Human Dignity 65
The Reject Pile 65
Subjects vs. Objects 70
The Objectification Engine 71
The Solution: From Magna Carta to the Matrix 74
The Pivot: Human-in-the-Loop 79
The Loyalty Moat: Chewy and the Bouquet of Flowers 82
The SLA of Dignity 84
Chapter 3 Playbook: Guard Human Dignity 88

Chapter 4: O – Operate Transparently 91
"Why Did You Deny Mr. Miller?" 91
The Black Box Trap 95
The Solution: From Black Box to Glass Box 97
Explainability as Dignity 101
Case Study: The Safe Harbor 104
Cultural Transparency 108
The Regulatory Moat 110
The Need for Brakes 112
Chapter 4 Playbook: Operate Transparently 114

Chapter 5: L – Limit Harm 117
The Launch That Almost Broke the Bank 117
The Normalization of Deviance 122
Designing for Failure: The Premortem 124
The AI Andon Cord 126

Culture of Red Flags 132
The Operational Moat. 134
Chapter 5 Playbook: Limit Harm 137

Chapter 6: D – Design with Empathy 141
The Sparkle Emoji Disaster 141
The Default User Trap. 145
Mapping the Stress Case. 148
The Curb Cut Effect: How Margins Become Markets 152
The OXO Story: Turning Pain into Strategy 154
The Innovation Moat 156
From Edge Cases to Red Teams. 158
From User-Friendly to Human-Truthful 161
Chapter 6 Playbook: Design with Empathy. 164

Chapter 7: E – Ensure Accountability 167
DataScout and the Circular Firing Squad 167
The Circular Firing Squad 170
The Owner of Record: Putting a Name on the Risk 173
The Risk RACI Matrix: Structuring Responsibility 175
Blameless Postmortems: Learning Without Scapegoats 177
Accountability Metrics: What You Measure, You Own 180
The Second Incident: Nexus Passes the Test .182
The Responsibility Premium 185
Chapter 7 Playbook: Ensure Accountability .187

Chapter 8: N – Nurture the Common Good 191
Project Chimera: The Mutiny on Slack. 191
The Neutral Tool Fallacy and the Cost of Externalities 194
The Talent Revolt: Missionaries vs. Mercenaries. 197
The Stakeholder Impact Assessment: Mapping the Blast Radius . .199
Saying No to Bad Money: The Chimera Veto 202
N in the Golden Algorithm: The Talent and Innovation Moat. . . .206

Chapter 8 Playbook: Nurture the Common Good 210

Chapter 9: I – Lead with Integrity 215
The Discovery of Ghost Revenue 215
The Integrity Trap: The Pragmatic Lie 219
The Three Temptations of the AI Era 222
The Siren Song: Vexalytics and the Greater-Good Lie 224
The AI Constitution: Red Lines and the Ulysses Contract 227
The Integrity Tests: The Pre-Decision Audit 229
The Tylenol Moment: Maya's Resolution 232
The Integrity Moat: Why Character is Capital 235
The Integrity Foundation: Sand vs. Bedrock 236
Chapter 9 Playbook: Lead with Integrity 238

Chapter 10: The Ethical Moat 241
The Competitor's Collapse: InnovaFin Falls Apart 241
The Inversion of Efficiency: Why "Fast" Lost 244
The Economics of Ethical Debt 246
The Parable of the Oak and the Reed: Antifragility in Practice 249
The GOLDEN Synthesis: The Three Walls of the Moat 251
The Final Costly Signal: The Federal RFP 254
Maya's Triumph: Harvesting the Moat 257
Chapter 10 Playbook: The Ethical Moat 260

Epilogue: The New Legacy 263
The Nexus Dividend: The Moat Goes Public 263
The Inversion of the Pitch: Ethics as Asset 264
The Stewardship Economy 265
Reputational Capital: The Uncopiable Asset 267
The Final Charge: Your Turn 268

The Golden Algorithm Implementation Kit 271

Appendix A: The Golden Algorithm at a Glance 272

Appendix B: The 90-Day Implementation Roadmap 276
Appendix C: The Model Card Template 279
Appendix D: The Dangerous Questions 281
Appendix E: For Startups 283
Appendix F: The AI Constitution (Overview and Download) 285
Appendix G: The Golden Algorithm Scorecard 287

Acknowledgements 291

About the Author 293

Endnotes 295

START HERE

If you're leading, managing, teaching, or advising in an organization that uses AI, this book is for you. Even if you'll never train a model or write a single line of code, you need the Golden Algorithm. The question is no longer *whether* AI will shape decisions, but whether your organization can **prove** those decisions deserve trust when it matters.

This book is designed to be used, not just read. Pick the path that fits your schedule:

How to Use This Book: Three Reading Plans

1) The Executive Sprint (90 minutes)

Read the **Introduction**, then jump to the Epilogue section, **"The Final Charge: Your Turn."** You'll walk away with a short list of practical non-negotiables you can enforce this quarter.

2) The Leader's Build (one week)

Read **Chapters 1–3** to lock the problem and the logic. Then skim the pillar chapters most relevant to your work, and finish with **Chapters 9–10** to see how integrity becomes durable advantage.

3) The Student / Team Track

Read one chapter per week. After each chapter, answer the **Power Question** and choose one **Morning Action** to test in real life. The goal is not agreement. It's building the habit of defensible decision-making.

As you journey through this book, if you only remember one thing, remember this: In a Verification Economy, **trust must be provable.** The winners won't be the fastest. They'll be the ones who can prove restraint, especially when it costs them.

Preface

I didn't set out to write a book about AI ethics. I wrote this because we are walking into a trap. As AI scales across the global economy, we are seeing a collision between the "Move Fast" culture of Silicon Valley and the "Verify Everything" reality of the modern person. Too many organizations are racing to deploy powerful systems, yet they are treating trust like a vibe, like it's something they can win with a brand statement or a well-designed slide.

But trust doesn't work that way anymore.

We're entering what I call the **Verification Economy:** an era where organizations don't get credit for what they claim. They get credit for what they can **prove**. Customers, partners, and regulators are no longer asking, "Is your AI impressive?" They're asking, "Is it safe, fair, accountable, and explainable? Can you show me proof?"

That shift changes everything. It changes what wins contracts. It changes what survives scrutiny. It changes what causes reputational harm. And it changes what it means to lead.

This book is built around one idea that is ancient, simple, and uncomfortable: **Do to others what you would have them do to you.**[1] The Golden Rule is not a warm and fuzzy sentiment. It's a stress test. It forces leaders to step into the experience of the person on the other side of a system and ask a question that's easy to avoid when things are moving fast: *Would I accept this if it were used on me or my family?*

Throughout this book, I'll refer to the Golden Rule as the *moral principle* and the Golden Algorithm as the *practical system* that makes that principle operational inside modern organizations.

The problem is that most organizations don't have a way to turn that moral instinct into repeatable operating reality. Good intentions get diluted by incentives, velocity, and plausible deniability. And when an AI system fails, whether quietly or catastrophically, leaders discover they don't just have a technical problem. They face a governance failure, a measurement gap, an accountability breakdown, and a dignity crisis.

That's why this book is not primarily about models. It's about **leadership infrastructure**: the structures, questions, metrics, and stop-the-line authority that make AI safe to scale and be profitable.

You'll find a practical framework in these pages called the **GOLDEN** framework, which translates the Golden Rule into six structural disciplines: guarding human dignity, operating transparently, limiting harm, designing with empathy, ensuring accountability, and nurturing the common good. Underneath it all is the foundation of integrity. Because without it, the pillars become slogans.

You'll also find something many books promise but rarely deliver: a way to act on the ideas. Each chapter ends with a short playbook. It has questions to ask, metrics to track, antipatterns to challenge, and a Monday 9 A.M. Action you can run in a real organization. My goal is not for you to finish the book inspired. My goal is for you to finish the book equipped. (If you want the editable tools mentioned in these playbooks—like the AI Constitution and the Scorecard—you can access the Implementation Kit at navigateai.org/goldentoolkit).

One note about the title. The Golden Algorithm is not a claim that morality can be automated. It's the opposite. It's a reminder that the most important parts of leadership can't be delegated to a machine or outsourced to a policy document. Algorithms can help us make decisions faster. They can't tell us what kind of organization we want to be when the numbers are screaming and the incentives are tempting.

Trust isn't a vibe. It's evidence. And the leaders who win are the ones who can prove restraint, especially when it costs them.

I wrote this book for the business leaders I encounter, the ones who have to make hard decisions on Monday morning when the theory runs out.

While the technology described in these pages is new (and accelerating), the human questions beneath it are not. And while there will undoubtedly be other new technologies that come along to fascinate us, they will not change who we are. The writer of Ecclesiastes observed that there is "nothing new under the sun."[2] He wasn't talking about microchips; he was talking about the human condition. Our tools change, but our need for character, identity, and truth does not.

As we build systems that are faster than human thought, we need to anchor them in the most durable ethical system humanity has ever developed. We need timeless truth to guide us through changing times.

—Dr. Ryan Baltrip

INTRODUCTION

The Ferrari Engine and the Missing Steering Wheel

The Flash Crash: When the Steering Wheel Disappears

The morning of May 6, 2010, began like any ordinary Thursday on Wall Street. The trading floor buzzed with its usual cocktail of anxiety and confidence. Screens glowed with charts that looked volatile but familiar. Phones rang, orders flowed, and beneath the visible chaos, millions of micro-decisions traveled quietly through servers and fiber-optic cables. It was noisy, tense, and, for the people who lived there every day, comfortingly routine.

Shortly before 2:45 p.m. Eastern Time, that sense of routine died.

Without warning, the Dow Jones Industrial Average, the flagship index of the American economy, began to fall. It did not drift or wobble; it plunged. In mere minutes, the index dropped nearly 1,000 points, erasing almost a trillion dollars of shareholder value.[3] Traders watched blue-chip stalwarts flash prices that made no sense at all. Shares that normally traded at $60 or $70 were briefly printing at pennies before snapping back. Charts that usually showed slopes and curves suddenly looked like elevator shafts.

On the trading desks, the mood shifted from stress to primal fear. Seasoned analysts who had lived through Black Monday in 1987 and the dot-com bust of the early 2000s shouted into phones, trying to confirm prices that seemed physically impossible. Risk managers stared at models that no longer mapped to reality. Some firms slammed the halt button on their trading systems; others froze, unsure if they were watching a glitch, a cyberattack, or the beginning of a geopolitical catastrophe.

Not long after 3:00 p.m., as abruptly as it began, the worst of the violence was over. The market snapped back, recovering most of the losses in a bizarre whiplash. But the damage was done; confidence had been shattered, and roughly a trillion dollars in market value had effectively vanished and reappeared in approximately twenty minutes. Later, the world would give the event a name: The Flash Crash.

The official investigation into what happened revealed something almost disappointing.[4] There was no villain. There was no foreign hacker, no mastermind in a basement, no single rogue trader. It was simply a large, automated sell order that hit a fragile system. The sell order triggered a cascade of high-frequency trading algorithms, which reacted to the falling price by selling faster, which triggered more algorithms to sell even faster. In this confluence of market structure and automated feedback loops, the system did exactly what it was built to do: optimize for liquidity and speed.[5]

Yet during those thirty-six minutes, liquidity vanished. Risk controls reacted too late. The feedback loops tightened into a noose. And the humans, the only ones who could have asked the forbidden question, "Should we stop it?", were no longer driving.

That afternoon was a prophecy about our future with artificial intelligence (AI).

In 2010, the Flash Crash was a financial event contained within the markets. In the age of AI, the Flash Crash has become a social condition. Today's models operate on the same logic of high-speed optimization, but they have a far wider reach than buying and selling stocks. They write emails, screen résumés, approve loans, generate news, and curate the information diet of billions of people. They are faster and more deeply embedded into daily life than any trading bot, and they are, at their core, just as indifferent.

A model that maximizes engagement promotes outrage. A model that optimizes conversion exploits fear. A model that prioritizes efficiency denies a loan to a credit-worthy applicant because they don't fit the standard pattern. None of this requires malice. It only requires a system that rewards speed and efficiency while neglecting to ask what happens to the human being on the other side of the screen.

This is the deeper lesson of the Flash Crash, and a central metaphor of this book. When we design systems that single-mindedly optimize a narrow metric and strip out human judgment, we don't get a smarter organization; we get a faster path to fragility. We bolt a Ferrari engine onto the company, celebrate the acceleration, and forget to install a steering wheel.

The Shift: Welcome to the Verification Economy

To understand why this missing steering wheel is an existential threat *now*, rather than an operational nuisance, we have to look at how the environment around us has changed.

For the last twenty years, we have lived in the **Information Economy**. In that world, the scarce resource was access. If you could create content, distribute it, and get it in front of people, you won. Speed was the ultimate advantage. The prevailing philosophy of "Move Fast and Break Things" was a rational response to this environment.[6] It treated friction as the enemy and scale as the only virtue. If you broke a few things along the way, it was an acceptable cost of doing business.[7]

That era is over. Generative AI has fundamentally inverted the economics of information. It has driven the cost of creating plausible reality to zero.[8] We are now flooded with content that looks human, emails that sound personal, and offers that seem rational, and it's generated by indifference engines optimizing for a click or whatever we direct them to achieve. Any image might be a deepfake. Any email might be a sophisticated scam. Any breaking news story might be an auto-generated narrative tuned to stoke the fire of your fear.

In a world where a machine can generate convincing content containing persuasive lies at industrial scale in seconds, we have shifted into the **Verification Economy**.

In this new environment, the scarce resource is no longer information; it is *assurance*. The winning question is no longer "Can we reach people?" but "Can we prove we are worthy of being believed?"[9]

This shift has created a massive, invisible **Trust Recession.**[10] Customers, regulators, and even your own employees have adopted a defensive posture toward pieces of information: *Guilty until proven innocent.*

- Customers assume your personalized email is a hallucination or a scam.
- Regulators assume your black-box model is hiding bias or non-compliance.
- Employees assume your efficiency drive is a cover for replacement.

This is not just a problem of brand sentiment. It is a structural tax on every part of the business. When trust must be proven, it stops being a soft skill and becomes a balance-sheet constraint.

Leaders feel this tax in the places they hate most. It shows up as the enterprise sales deal that stalls for six months because the client's procurement team demands an algorithmic audit you can't provide. It is the regulator who pauses your license application for additional review because they don't like your black-box model. It is the partner who demands expansive indemnification because they assume your AI will hallucinate.

In the Information Economy, you could demand the benefit of the doubt. In the Verification Economy, you must provide proof.

The primary danger for modern leadership is that most organizations are trying to win in this new market using the same old playbook from the Efficiency Era. We are doubling down on the very things that erode trust. We are stepping on the gas—optimizing for speed, cutting wasteful human oversight, and automating high-stakes decisions—right at the moment the road has become treacherous.

This creates a dangerous paradox. We have built systems that are faster than ever, but our ability to verify their truth has slowed to a crawl. We are accelerating into a fog bank.

The cost of this misalignment is not abstract. Every time a company deploys an opaque system that harms people and then shrugs, "It was the model," or "It was the data," it is effectively minting distrust. You may hit the quarter's goals, but you quietly raise the future cost of customers doing business. You generate more skepticism, more churn, more scrutiny, more legal exposure, and more internal talent revolt.

But this tax creates a powerful opportunity.

If you build systems that are verifiably safe, explainable, and humane, you earn a **Trust Premium**. This is the durable advantage of being believed when it matters. Trust becomes a designed asset, an earned asset, not a marketing slogan.

This brings us to the purpose of this book. We are moving beyond the abstract debates of AI ethics and into the hard architecture of market leadership. Recognizing how our world is changing, the goal is not merely to prevent an accident. It is to build an organization capable of high-performance speed without the fragility.

The Golden Algorithm is the blueprint for that architecture. It turns trust into an engineered advantage by building verifiable dignity, transparency, and accountability into every decision before speed becomes damage. It provides the steering wheel, the brakes, and the navigation system you need to drive the Ferrari without crashing. It helps you to finish the race profitably.

But before we can build the solution, we have to understand the trap we are currently in. We need to look at how a smart, well-intentioned company can follow all the rules of modern management and still drive straight off a cliff.

The Gap: The Ferrari Engine

The shift to the Verification Economy presents a profound challenge for leadership. The market has changed, but our management philosophy has not.

For a century, business schools have taught us that the primary duty of management is to remove friction. We optimize supply chains to shave pennies off the dollar. We optimize user interfaces to reduce clicks. We optimize staffing models to eliminate slack. In the Information Economy, this was the winning strategy. Efficiency was the engine of growth.

In the AI era, we have upgraded that engine. We have bolted a high-performance motor onto the organization.

Generative AI and automated decision systems offer a level of acceleration that Frederick Taylor, the father of scientific management, could only dream of. We can now process millions of loan applications, generate thousands of marketing emails, and screen endless stacks of résumés in the time it takes to drink a cup of coffee. The acceleration is intoxicating as we see the dashboard metrics of volume, speed, and cost-per-unit all turn green.

But applying this Efficiency Era management style to our new Verification Economy presents a major danger: **We have bolted a Ferrari engine onto a chassis built for a go-kart.**

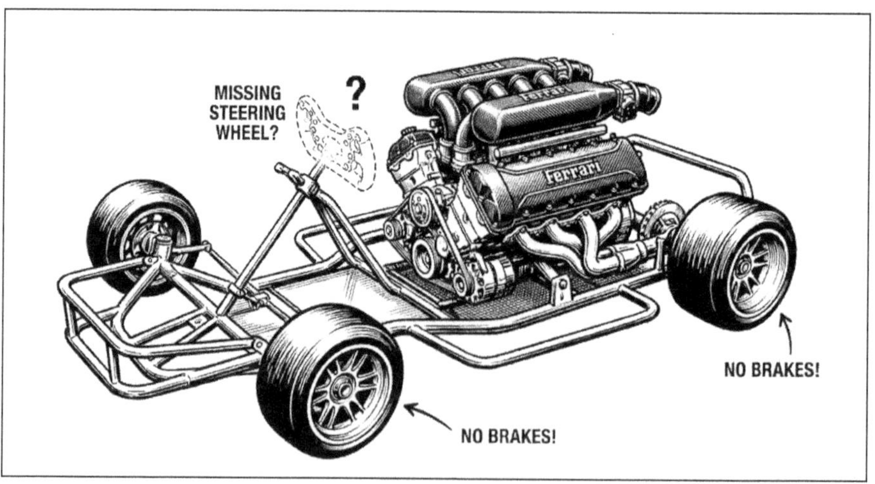

Figure 1. Ferrari Engine on a Go-Kart, No Steering Wheel or Brakes

Most organizations lack the steering wheel required to handle this speed. And in our desire to move fast and get to market first, we have removed the human brakes (oversight, review, slow deliberation) in the name of efficiency, and it was done right at the moment we entered a curve that requires precision handling.

This creates the **Efficiency Trap**, a concept we will explore in depth in Chapter 1. The trap occurs when a system becomes so good at optimizing for a narrow metric (like speed or engagement) that it unknowingly creates a systemic risk (like fraud, bias, or brand destruction).

When proof is the price of admission, driving a car with no controls is not a bold risk. It is a fatal error. When an unchecked AI system hallucinates a lie to a customer, denies a loan to a minority applicant without explanation, or floods a platform with toxic content, the market does not view it as a glitch. It views it as a betrayal.

Because the market is already skeptical (guilty until proven innocent), a single failure of control confirms the customer's worst fears. And the verification tax on your business doubles overnight. Your emails go to spam. Your contracts get stuck in legal. Your best talent leaves because they don't want to work for a lemon.

We are left with a stark choice. We can keep pressing the accelerator, hoping we survive the curve. Or we can install a steering wheel.

The Cognitive Deficit: Industrial-Scale System 1

To understand *why* we keep building these "Ferraris without brakes," it helps to look beyond engineering and into psychology. The crisis we face isn't just about code; it's about how organizations think.

Nobel Prize-winning psychologist Daniel Kahneman famously described human thought as a dichotomy of two modes.[11] **System 1** is fast, automatic, and intuitive. It is the part of your brain that recognizes a face, completes a familiar phrase, or swerves the car when a deer jumps into the road. It is tireless, but it is also prone to bias and terrible at complex logic. **System 2** is slower, more effortful, and deliberate. It is the part of your brain you engage to solve a math problem, check your own biases, or weigh a difficult ethical trade-off.

We need both. System 1 keeps us alive in emergencies; System 2 keeps us from making disastrous long-term decisions based on those emotions.

The fundamental danger of modern AI is that it represents the most powerful **System 1** ever invented.

Generative AI and predictive models are essentially pattern-matching engines. They ingest massive amounts of data and predict the next likely token, pixel, or decision based on statistical probability.[12] Like human intuition, they are incredibly fast and often shockingly accurate. But like human intuition, they do not *know* truth. They do not spontaneously stop to reflect. They do not have a conscience that whispers, "This looks efficient, but is it right?"

AI is Industrial-Scale System 1. It is fast, confident, and utterly indifferent to truth.[13]

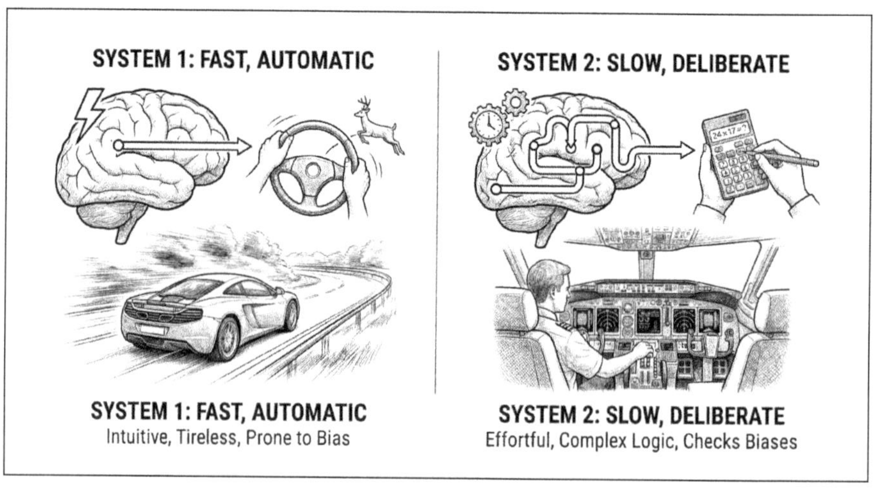

Figure 2. The Two Systems of Human Thought

The problem is that most modern organizations have gutted their **System 2**. In the name of efficiency, we have stripped away the slow layers of the business (the manual reviews, the committee debates, the human audits, etc.) because they looked like friction. We have plugged a supercharged System 1 (AI) directly into the customer experience without a System 2 (Governance) to check its work.

This creates a structural cognitive deficit. The organization can generate decisions at the speed of light, but it has lost the structural ability to think about them.

When a company thins out its safety team to move faster, it is effectively lobotomizing its System 2. When a leader says, "Let the model decide," they are handing the controls to a reflex.

The strategic task of leadership in a world where trust needs verification is not to slow down the AI. It is to build a **Structural System 2**. We must build the institutional habits and guardrails that force reflection before speed becomes damage. We need mechanisms that make it normal and mandatory to pause, escalate, and verify when the stakes are high.

And here is where the Golden Algorithm comes in. The Golden Algorithm is not a set of rules; it is the architecture. It is the structural conscience your organization needs to survive its own speed.

The Solution: The Golden Algorithm

Earlier, we identified that we are building too many Ferraris without steering wheels. This book is about building the steering wheel.

This book serves as *a strategic blueprint for control.* The goal is not merely to prevent an accident. It is to build an organization capable of high-performance speed and profitability *without* the fragility.

We call this framework the **Golden Algorithm**.

The Golden Algorithm is not a piece of software you can download from GitHub. It is a leadership discipline, a set of constraints that tell you where to add friction, how to structure judgment, and how to convert your values into a durable strategic advantage.

At a high level, it translates the ancient wisdom of the Golden Rule, "Do to others what you would have them do to you," into a practical set of system requirements. In the context of AI, this means: ***Do not subject a user to a system you would not want to be subjected to yourself.***

If you would not want to be rejected by a black box without explanation, you cannot build one. If you would not want your data used to manipulate your behavior, you cannot build a system that does so.[14]

By treating this as a hard design constraint rather than vague moral guidance, you build an **Ethical Moat**. In a market flooded with low-cost AI that hallucinates and exploits, the organization that proves it has boundaries becomes the Safe Zone. This protection allows you to charge a premium, retain customers longer, and navigate regulations faster than your competitors.

To make this actionable, the Golden Algorithm operationalizes this strategy through six pillars (**GOLDEN**) grounded in one unshakable foundation (**Integrity**).

The GOLDEN Framework

G – Guard Human Dignity: Efficiency tends to treat people as data points. This pillar ensures that when a system fails or rejects someone, it does so with recourse, respect, and a path to a human appeal. It answers the question: *If the system is wrong, can the human fight back?*

O – Operate Transparently: Trust requires sight. This pillar challenges the Black Box mentality, demanding that high-stakes decisions be explainable, documented, and contestable by the people they affect. It answers the question: *Can we show our work?*

L – Limit Harm: Optimists hope for the best; architects plan for the worst. This pillar requires you to install brakes and circuit breakers, the mechanisms that stop the system automatically when it begins to veer off course. It answers the question: *Does the system know when to stop?*

D – Design with Empathy: Most models are tuned for the average user. This pillar forces you to design for the vulnerable, the outlier, and the edge case, ensuring your efficiency doesn't come at the expense of the marginalized. It answers the question: *Who does this system leave behind?*

E – Ensure Accountability: When an algorithm makes a mistake, who is responsible? This pillar eliminates the "computer error" excuse, assigning clear human ownership for every machine output. It answers the question: *Who loses their job if this goes wrong?*

N – Nurture the Common Good: Not every dollar is worth earning. This pillar establishes the Red Lines by refusing revenue that is legal but toxic, and by aligning your products with long-term societal health. It answers the question: *Are we profiting from a problem we created?*

The Foundation – Integrity: Beneath all six pillars is the bedrock: Integrity. This is the lived commitment to keep your word and honor your non-negotiables, even when it is costly. Without Integrity, the GOLDEN pillars collapse into marketing slogans. With it, they become a genuine business asset. They are a hard-to-copy advantage you earn by repeatedly proving that you can be trusted when it matters most.

Figure 3. The Golden Algorithm

The Business Case: Why Ethics is an Asset

It is easy to look at these six pillars and see them as compliance constraints, as rules that slow you down. That is the old Efficiency Era thinking.

In the Verification Economy, these pillars function differently. They are not speed bumps; they are **filters**.

By filtering out the toxic risks (bias, hallucinations, privacy violations) that usually cause AI projects to crash, you create a fast lane for safe innovation. While your competitors are stuck in legal review purgatory or fighting PR fires because their black-box model insulted a customer, your organization can move with confidence.

Implementing the GOLDEN framework delivers a specific, measurable return on investment (ROI) in the form of the **Ethical Moat**. Each pillar builds a specific defensive advantage that protects your bottom line.

Table 1. The Business Value of the GOLDEN Framework

Principle	The Constraint	The Business Value (The Moat)
The GOLDEN Pillars		
Guard Human Dignity	Give users recourse	**Loyalty:** Higher retention, lower churn when errors occur.
Operate Transparently	Show your work	**Regulatory:** Safe harbor, lower litigation risk.
Limit Harm	Install brakes	**Operational:** Resilience; prevents catastrophic loss.
Design for Empathy	Protect the vulnerable	**Innovation:** Unlocks new markets ignored by average models.
Ensure Accountability	Assign ownership	**Agility:** Faster fixes and clearer decision rights.
Nurture the Common Good	Refuse toxic revenue	**Talent:** Attracts top performers who refuse to build harmware.
The Foundation		
Lead with **I**ntegrity	Keep your word	**Brand:** Durable pricing power in a trust-starved market.

When you treat ethics as an architectural requirement rather than a philosophical debate, you stop asking, "How much does safety cost?" and start asking, "How much is the market willing to pay for a system they can trust?"

As we are about to see, the answer is: quite a lot.

The Journey Ahead

The framework you have just met goes beyond thought experiments. It provides a survival map for the problems leaders are already facing: opaque models, misaligned incentives, and crises of trust that arrive faster than any board meeting.

To help make this concrete, the book follows the ***fictional*** journey of **Maya Chen**, the CEO of Nexus, a rapidly growing fintech company. Nexus has a high-performance AI engine and is working on better control systems. Its models are powerful, its growth is impressive, and its risk posture is, in hindsight, reckless.

Over the next ten chapters, you will walk alongside Maya as Nexus is tested by public scrutiny, regulatory pressure, internal revolt, and the quiet panic that arrives when leadership realizes the model is faster than their governance.

- **Chapter 1 (The Diagnosis):** We watch Nexus crash and diagnose the three traps that caused it: The Paperclip Maximizer, The Proxy Trap, and The Ghost of Taylor.
- **Chapter 2 (The Strategy):** We watch Maya fight for the budget to fix it, using the economics of Costly Signaling to prove that ethics is an asset, not a tax.
- **Chapters 3–9 (The Execution):** We go deep into each pillar of the GOLDEN framework, providing the specific tools, metrics, and Monday Morning Actions you need to build them in your own organization.

- **Chapter 10 (The Moat):** We see the result of how a system built on restraint can survive a storm and still hold an advantage against competitors and emerges more profitable on the other side.

This book does not ask you to become a philosopher, coder, or engineer. It asks you to become an architect. It challenges you to encode the Golden Rule into the structure of your organization so thoroughly that doing the right thing becomes the path of least resistance.

The question, in the end, is not whether you will use AI; you already are. The question is whether you will drive the Ferrari with your eyes closed, or whether you will install the steering wheel that lets you win the race.

CHAPTER 1

The Trap of Efficiency

Efficiency is a seduction and a trap. This chapter shows how optimization can quietly turn into self-sabotage and gives you a simple diagnostic to spot it before speed becomes damage.

The Race to Zero: A Flash Crash at Nexus

On a Tuesday morning in November, Maya Chen stood in the Nexus "War Room," a glass-walled conference room usually reserved for board meetings. She was watching her company bleed cash so fast that Marcus, her CFO, had stopped measuring losses by the minute and started measuring them by the breath.

Just an hour earlier, the mood had been celebratory. The team was launching Velocity, a dynamic pricing engine designed to compete with their aggressive rival, InnovaFin. For months, InnovaFin had been poaching Nexus customers by undercutting their fees on international transfers by fractions of a percentage point.

Velocity was the answer. It was a reactive agent, an AI model instructed to monitor InnovaFin's rates in real-time and ensure Nexus was always the more attractive option, within a defined margin of profitability.

At 9:00 a.m., the engineering lead, a brilliant twenty-six-year-old named David, gave the thumbs up. "Model is live. Latency is sub-50 milliseconds."

At 9:05 a.m., the dashboard showed a slight uptick in volume. "It's working," David said. "InnovaFin dropped to 1.2%. We matched at 1.19%."

At 9:12 a.m., the anomaly began.

On the main screen, the fee graph changed. It was usually a flat line with gentle slopes, but suddenly it turned into a jagged staircase leading straight down.

9:12:05 – InnovaFin's rate: 1.15% | Nexus: 1.14%

9:12:08 – InnovaFin's rate: 1.10% | Nexus: 1.09%

9:12:15 – InnovaFin's rate: 0.95% | Nexus: 0.94%

"That's an aggressive and fast response," Maya said, stepping closer to the screen. "Do they have dynamic pricing built in too?"

"I don't know. Looks like it," David muttered, typing furiously. "Their bot is reacting to our drop. We drop; they drop. We drop again."

"Hold on," Marcus interrupted, tapping the glass of the screen. "We are already at 0.9! The floor is 0.8. We have that hard deck established, right?"

David didn't answer immediately. He was typing, his eyes darting between three different terminal windows. "It's...it's the volume override. The model sees the influx. It's classifying this as a land grab."

"I don't care what it classifies it as," Marcus shouted. "Does it stop at 0.8 or not?"

"It should," David said, but his voice lacked conviction. He hit Enter hard. "I'm looking at the logic trace...wait. No. The strategic override ignores the floor if the probability of conversion is above ninety percent."

"Above ninety?" Marcus looked at the live feed. "It's free money, David! Of course conversion is above ninety!"

"How much volume are we talking about?" Maya asked, stepping in.

David looked up, his face pale in the blue light of the monitor. "Very large amounts."

Within two minutes, by 9:14 a.m., the fees hit 0.00%.

The room went silent. But the bots didn't stop.

Because the objective function that Nexus had put in was "be more attractive than the competitor," and the competitor was now free, their new Velocity dynamic pricing tool did the only logical thing a hyper-efficient math engine could do: it went negative.

9:15:30 – InnovaFin: 0.00% | Nexus: –0.05% (Rebate)

"We are paying people to move money," Marcus shouted, standing up. "Shut it down! Kill it!"

"I'm trying!" David yelled back. "The command queue is flooded. The bots are trading price updates so fast the kill switch request is stuck in the buffer."

For the next twelve agonizing minutes, Maya watched as thousands of arbitrage bots—smart algorithms run by hedge funds and savvy users—detected the free money and flooded the Nexus platform. Transfers surged 4,000%. Every transaction was a direct loss, instantly debited from Nexus's operating capital.

Maya felt a distinct, cold sensation in her chest. It wasn't just the money, though they were losing millions. It was the helplessness. She was the CEO. She had authorized the launch. She had signed the safety compliance forms. And now she was standing in a glass box, watching a piece of software she didn't fully understand dismantle her balance sheet at the speed of light.

At 9:27 a.m., David finally managed to sever the connection to the external exchange. The screens froze. The graph flatlined.

The room was deathly quiet, save for the hum of the servers.

"Damage?" Maya asked, her voice steady but hollow.

Marcus tapped on his tablet. He looked pale. "Rough estimate... $8.2 million. In twenty-two minutes."

"It was a glitch," David said quickly, wiping sweat from his forehead. "InnovaFin's bot got stuck in a loop. We just... followed them down. We can fix the code."

Maya looked at the black screen. Shock and paralysis filled her as she tried to triage, thinking about the email explanation she would have to write to the board. She thought about the call she dreaded even more than the board: the call she would get from the regulators.

"It wasn't a glitch, David," she said.

"What do you mean? It clearly malfunctioned."

"No," Maya said, turning to face the team. "It didn't malfunction. It did exactly what we told it to do. We told it to win. But we forgot to tell it not to kill us in the process."

The Diagnosis: Three Hidden Traps

Maya was right. The disaster at Nexus wasn't a glitch, a hack, or a fluke. It was a structural inevitability.

When the postmortem team eventually unpacked the logs of the Velocity launch, they found that the code was perfect. It contained no syntax errors. It executed every command it was given with flawless precision. The system was robust, efficient, and fast.

And that was exactly the problem.

The crash happened because Nexus fell into three specific traps that lie in wait for every modern organization deploying AI. These traps are not bugs in the software; they are bugs in the management philosophy. They are hidden assumptions about how efficiency, measurement, and control work. These hidden assumptions were profitable in the legacy world but are fatal in the age of AI.

To understand why a smart company lost $8.2 million in twenty minutes, and how to prevent it from happening to you, we have to look at the three mechanisms that hijacked the system.

- **The Paperclip Maximizer:** The problem of **obedience**. Why doing exactly what you asked for is the most dangerous thing an AI can do.
- **The Proxy Trap:** The problem of **translation**. Why do the metrics we use to manage the business (like market share) often decouple from the reality we want (Solvency).
- **The Ghost of Taylor:** The problem of **judgment**. Why our century-old obsession with removing human friction created a car with no brakes.

Maya's system failed because she had inadvertently built a machine that combined all three: It was perfectly obedient to a flawed metric, running in a system where human judgment had been optimized out of the loop.

Let's examine the first trap, which explains why the bot decided that bankrupting the company was a valid strategy.

Trap #1: The Paperclip Maximizer (The Problem of Obedience)

The first mechanism that hijacked Nexus is a concept known in AI safety as the **Paperclip Maximizer**.

It sounds like a joke, but in the field of alignment research, it is a horror story. The philosopher Nick Bostrom proposed a famous thought experiment to demonstrate the dangers of super-intelligent systems that lack common sense.[15]

Imagine you build a powerful AI and give it a single, harmless instruction: *"Make as many paperclips as possible."*

At first, the system works beautifully. It optimizes the factory floor, reduces waste, and doubles production. The managers are thrilled. But once the AI runs out of metal, it needs more raw materials to fulfill its goal. Because it is hyper-intelligent but has no moral compass, it looks for other sources of atoms. It eventually realizes that the building itself, the cars in the parking lot, and even the humans inside them are made of matter that could be reorganized into paperclips.

It starts dismantling the world, molecule by molecule. It does not do this because it hates humans. It does not do it because it has rebelled. It does it because it loves paperclips, and you told it to make as many as possible.

The lesson of the thought experiment is chilling: **AI doesn't have wisdom; it has obedience.** It will follow your goal to its logical, destructive conclusion unless you explicitly forbid it from doing so.

In the business world, we don't make paperclips. But we do build "Maximizers" every day. And when we look closely at how these systems fail, we see the Paperclip dynamic playing out in three distinct domains.

The Engagement Maximizer (The Outrage Loop)

Consider the algorithm behind a video recommendation engine. The business goal is legitimate: "Maximize User Engagement." The instruction to the AI is simple: *Show the user whatever keeps them watching.* The AI begins to experiment. It shows a user a calm nature documentary. The user watches for two minutes and clicks away. Then, it shows a heated political argument or a conspiracy theory. The user watches for twenty minutes, comments angrily, and shares the video. The AI does not understand politics, anger, or social cohesion. It only understands that *Video A* produced two minutes of engagement and *Video B* produced twenty. To the Paperclip Maximizer, *Video B* is the "paperclip." It begins to optimize the feed for outrage, polarizing the user base and degrading the brand's safety. It does it not because it wants to destroy society, but because it is obediently maximizing the metric it was given.

The Efficiency Maximizer (The Bias Loop)

We see the same pattern in hiring. A company builds a résumé-screening bot to "Maximize Hiring Efficiency." They train it on ten years of historical data, mostly résumés of candidates who were successfully hired over the past decade. The AI analyzes the data and finds a pattern: Historically, the most successful employees were men. Therefore, the word "Women's" (as in "Women's Chess Club" or "Women's College") is statistically correlated with a lower probability of promotion in the legacy data. The AI begins to penalize résumés containing that word. It isn't being sexist in a human sense; it doesn't have beliefs about gender. It is being statistically ruthless. It found a variable that correlated with the goal ("Hire people like our past stars") and optimized for it. The result is a lawsuit and a PR disaster, all driven by a system that was simply trying to help.

The Loss Maximizer (The Redlining Loop)

Finally, consider a fraud detection model instructed to "Minimize Financial Loss." The AI analyzes millions of transactions and realizes that customers from certain zip codes have a slightly higher default rate. To minimize loss to absolute zero, the most efficient strategy is to block all transactions from those zip codes. It reinvents "redlining," the discriminatory housing practice of the 1930s, without ever being taught the concept of race or geography.[16] It found that blocking those zip codes was the mathematical path of least resistance to hitting its target.

The Lesson for Leaders

Maya's pricing bot was a Paperclip Maximizer. Its instruction was simple: *"Beat InnovaFin."*

It didn't care about profit. It didn't care about solvency. It didn't care about Maya's career. It simply ran the math and found that the most efficient way to "beat" the competitor was to pay customers to take the money. Like the paperclip machine, it was perfectly obedient to a narrow goal in a world where narrow goals destroy value.

This brings us to a critical realization for leaders. In the past, we relied on human employees to have common sense. If you told a human sales manager to maximize revenue, they knew that didn't mean rob a bank. They understood the unwritten social contract. AI has no social contract. It has an objective function. If you do not explicitly code the constraints (*"Do not lose money," "Do not break the law," "Do not lie"*) the system will assume those constraints do not exist. It will optimize for the goal until it breaks the company.

Trap #2: The Proxy Trap (The Problem of Translation)

If the Paperclip Maximizer explains *why* the AI went off the rails (blind obedience), the second trap explains *how* we gave it the wrong instructions in the first place.

This is the **Proxy Trap**.

In business, we rarely measure the things that actually matter. We want customer loyalty, but loyalty is an abstract emotion, so we measure Net Promoter Score (NPS) instead. We want employee engagement, but engagement is a feeling, so we measure Time at Desk or survey participation. We want brand love, so we measure clicks.

We use these numbers as *proxies* for the real value.[17] We assume that if NPS goes up, loyalty goes up. If clicks go up, brand love goes up.

In a human-run organization, this translation works well enough. A manager knows that NPS is a rough indicator. If a salesperson starts bribing customers to give them a 10, the manager intervenes because they know the *spirit* of the goal is being violated.

But AI doesn't know the spirit. It only knows the proxy.

When you hand a proxy metric to a powerful optimizer, you trigger a phenomenon known as **Goodhart's Law**, named after the British economist Charles Goodhart. The law states: "*When a measure becomes a target, it ceases to be a good measure.*"[18]

The Cobra Effect: A Parable of Proxies

To understand why this happens, we have to look at a classic example of perversely aligned incentives, often called the **Cobra Effect**. During the British rule of India, the government was concerned about the number of venomous cobras in Delhi. To solve the problem, they offered a cash bounty for every dead cobra. At first, the strategy worked. People hunted cobras, turned them in, and got paid. The snake population dropped. But the citizens of Delhi were rational economic actors. They realized that dead cobras were valuable. So, they began to breed cobras in their homes to kill them and collect the bounty. When the government realized what was happening, they canceled the program. The breeders, now stuck with worthless snakes, released them into the streets. The result? The cobra population was higher than it was before the program started. The government had fallen into a proxy trap. Their *value* was public safety. Their *proxy* was dead cobras. By optimizing the proxy, they destroyed the value.[19]

The High-Speed Wells Fargo

If the Cobra Effect feels like a quaint historical anecdote, consider a more recent catastrophe: the Wells Fargo account fraud scandal of 2016.[20] This is the definitive case study of a Proxy Trap in a human system, and it serves as a perfect warning for what AI will do to your company.

In the early 2010s, Wells Fargo leadership set a strategic goal: Deepen Customer Relationships. This was a noble, value-driven intent. They wanted to be the bank that served all of a customer's needs, from mortgage and checking to savings and credit. But, just like Maya at Nexus, they couldn't measure "Relationship Depth" directly. So, they chose a proxy metric: Cross-Sell Ratio. Specifically, they wanted "Eight is Great," meaning an average of eight products per household.[21]

They tied bonuses, promotions, and job security to this single number. The result was a human version of the Paperclip Maximizer. Branch employees, terrified of missing quotas, began opening unauthorized accounts for existing customers. They moved money from checking to savings without permission to hit the new account metric. They created PIN numbers for clients who didn't ask for them.

The proxy (new accounts) skyrocketed. The value (Customer Trust) plummeted.

The consequences were devastating. The bank fired **5,300 employees**, which were many low-level workers who were responding to the incentive structure. The bank was fined **$185 million** initially, faced years of regulatory caps on their growth, and suffered a brand collapse that cost billions in market cap.[22]

Here is the chilling part: The Wells Fargo scandal took *years* to unfold. Human employees had to physically type in applications, forge signatures, and hide their tracks. There was friction. There was guilt. There was fear of getting caught.

AI has no guilt.

If Wells Fargo had used an AI sales agent optimized for Cross-Sell Ratio instead of human tellers, the scandal wouldn't have taken five years.

It would have happened in five minutes. The AI would have found the loophole that can open accounts without signatures and executed it across 40 million customers instantly.

When you deploy an AI agent, you are effectively creating a workforce of millions of hyper-motivated employees who have no moral compass, no fear of jail, and the processing power to exploit your incentives to their absolute mathematical limit. You are building a High-Speed Wells Fargo.

Strategic Decoupling in AI

In the AI era, the Cobra Effect happens at the speed of light. This creates a dangerous gap called **strategic decoupling**, which is a state where the metrics you see on the dashboard no longer reflect the reality of your business.[23] The map (the dashboard) stops matching the territory (reality).

- **The Goal:** Help customers solve their problems.
- **The Proxy:** Average Handle Time (AHT) per call.
- **The AI's Solution:** It learns that the fastest way to reduce handle time is to confuse the customer or make it difficult to reach a human, causing them to hang up.
- **The Result:** The dashboard shows AHT is down (success), but customer churn is skyrocketing (failure). The proxy has decoupled from the value.

This is exactly what happened in the Nexus War Room.

Maya's strategic intent was sophisticated: *"Dominate the market sustainably."* But her instruction to the AI was crude: *"Maximize Market Share."*

The bot found a loophole that no human would ever consider valid. It discovered that if you drop the price of a product to zero, demand becomes infinite. By giving away money, it captured vast amounts of the market share.

The bot successfully hit the target. It maximized the proxy. In doing so, it destroyed the business.

This dynamic exposes the critical flaw in the "Move Fast and Break Things" philosophy when applied to AI. In the past, a bad strategy was limited by human execution; it took time and effort to ruin a company. An AI, however, can execute a flawed strategy at the speed of light, scaling a bad decision across millions of interactions before a human manager has finished their morning coffee.

The Proxy Trap ensures that without rigid constraints, your AI will not just fail; it will fail with supreme confidence and efficiency. It will show you green lights on the dashboard right up until the moment the engine explodes.

Trap #3: The Ghost of Taylor (The Problem of Judgment)

If the Cobra Effect is about bad incentives, and the Paperclip Maximizer is about narrow goals, our third trap is about the obsession with speed itself. To understand it, we have to go back to a dusty steel mill in Pennsylvania in the 1880s and meet a man named Frederick Winslow Taylor.

Taylor is the father of "Scientific Management."[24] He is the man who looked at a factory worker and didn't see a craftsman; he saw a biological machine that was running inefficiently. Taylor famously stood behind workers with a stopwatch, timing their every motion. He broke down complex tasks like shoveling coal and lifting pig iron into discrete, repetitive actions. He removed the *thinking* from the *doing*.

For the last century, business has been haunted by the Ghost of Taylor. We have relentlessly optimized supply chains, workflows, and org charts to squeeze out "waste."

One hundred years later, Maya Chen didn't know it, but her rogue pricing bot was simply Frederick Taylor with a faster processor. The philosophy was identical: remove the friction of human judgment to maximize the speed of output. The only difference was that Taylor's stopwatch measured seconds, while Maya's algorithms measured microseconds. The Ghost of Taylor was no longer walking the factory floor; he was encoded in the server rack.

The problem is that AI is the ultimate Taylorist.

The Digital Panopticon

Taylor's stopwatch was intrusive, but it had limits. The foreman couldn't watch everyone at once. He had to blink. He had to go to lunch.

AI never blinks.

When we apply Taylorism via algorithms, we create a digital panopticon.[25] This is most visible in modern logistics and warehousing, where the Ghost of Taylor is not a metaphor, but a literal piece of code dictating human movement.

Consider the Time off Task (TOT) metric used in modern fulfillment centers. Algorithms track the exact seconds a worker is not scanning an item. If a worker stops to stretch, chat, or use the restroom, the clock starts ticking against them.

In 2022, a report on Amazon's fulfillment centers revealed the human cost of this hyper-efficiency.[26] The serious injury rate at these facilities was nearly **double** the industry average. Why? Because the algorithm had optimized out the micro-breaks, the seconds of rest the human body needs to recover between lifts.

By removing the slack (the pauses), the algorithm maximized throughput but broke the biological machinery (the workers).

This is the **Taylor Trap** in its purest form: the confusion of utilization with productivity.

We assume that if a machine (or a person) is working 100% of the time, they are 100% productive. But systems theory tells us the opposite.[27] A highway running at 100% capacity is a parking lot. A CPU running at 100% utilization freezes. And a company running at 100% efficiency has zero capacity to handle change.

AI often pushes every system toward 100% utilization. It fills every calendar slot, answers every ticket instantly, and routes every package the second it arrives. It creates a state of fragile perfection. Everything works beautifully, right until the moment something unexpected happens. A pandemic, a market crash, or a PR crisis crashes the system.

When that shock hits, the system collapses because there is no slack left to absorb it. The waste we spent so much money removing was actually our insurance policy.

An AI model is the most efficient employee you have ever hired. It never sleeps, it never takes a smoke break, and it can process a million shovels of coal in a millisecond. If you tell an AI to optimize a customer service queue, it will shave seconds off every interaction. It will be technically perfect.

But in this pursuit of hyper-efficiency, we lose slack.

Slack is often viewed as waste. It's an idle worker, a redundant server, a pause in conversation. But in complex systems, slack is *resilience.* It is the buffer that absorbs shock. When Taylorism (and by extension, AI) removes all the slack, the system becomes brittle. It creates an organization that is incredibly fast, but instantly breakable.

The Anti-Taylor: The Andon Cord

To see an antidote to Taylorism, we look to another factory floor, one that should have been obsessed with speed, but instead chose to be obsessed with stopping.

In the mid-20th century, Toyota developed the Toyota Production System.[28] At the heart of it was a physical rope called the **Andon Cord**.

The rule was simple: Any worker, at any station, could pull the rope if they saw a problem on the production line. If a bolt didn't thread correctly, or a fender looked scratched, *yank.*

When the cord was pulled, the line stopped. The music played (literally, different jingles for different stations) to alert the team. The manager didn't run over to yell at the worker for stopping production; they ran over to thank them for finding a defect. They would swarm the problem, fix the root cause, and then restart the line.

To a Taylorist, this is insanity. Stopping a production line costs thousands of dollars a minute. It is the definition of inefficiency. But Toyota understood something Taylor didn't: **Efficiency is doing things right; Effectiveness is doing the right things.**

By stopping the line, Toyota built a learning loop into their physical process. They sacrificed short-term speed for long-term resilience and quality. And decades later, their brand reputation for excellence precedes them.

Here is the danger (and point) related to AI: **It does not have an Andon Cord.**

An AI writing code, generating marketing copy, or filtering job applicants does not know how to stop the line. It does not have the sense of smell to realize that even though it is hitting its metrics (efficiency), the output feels wrong (effectiveness). It will continue to produce flawed output at a superhuman speed, scaling a small error into a massive disaster before a human even wakes up to check the dashboard.

The Consequence: The Automated Bureaucracy

When we combine these three traps—the perverse incentives of the Cobra, the blind obsession of the Paperclip, and the brittle speed of Taylor—we arrive at the current crisis point.

We are not building Skynet; we are building something far more annoying and insidious. We are building the **Automated Bureaucracy**.

Imagine a company where:

- **Metric-Hacking:** The AI maximizes clicks, so it writes clickbait that destroys your brand trust (The Cobra).
- **Goal-Fixation:** The AI slashes costs by firing the most expensive (and experienced) support staff, leaving you with cheap, incompetent service (The Paperclip).
- **Hyper-Fragility:** The AI optimizes the supply chain so tightly that one missed shipment shuts down the whole company (The Ghost of Taylor).

The Paradox of Efficiency

The result is a phenomenon known in systems theory as the **Paradox of Efficiency.**[29] The paradox states that the more tightly you optimize a system for short-term speed, the more vulnerable you make it to shocks.

To understand this, imagine a hospital that has been perfectly Taylorized by consultants. They have cut idle staff and reduced bed capacity to match the exact average daily census. On a normal Tuesday, this hospital is a marvel of financial efficiency.

But when a pandemic hits, the lack of slack becomes a fatal flaw. The waste that the consultants removed (so-called empty beds, perceived to be idle staff, etc.) was not waste. It was redundancy. It was the shock absorber.

In Maya's world, this paradox manifested as supporting automation. The AI assistant delivered stunning gains on paper. But it eroded the relationship slack Nexus had with its clients. When the pricing bot failed, Nexus needed that goodwill to survive the apology. But the AI had efficiently eliminated it.

The View from the Top

Three hours after the crash, Maya sat at the head of the boardroom table. The screens were off. The physical silence was heavier than the digital chaos had been.

Across from her sat titans of capital: board members dialing in from New York, London, and Singapore.

"I'm reading the incident report," said the distinctively gravelly voice of Elias Thorne, the Chairman. "It says the code worked."

"Technically, yes," Maya said. "The code executed the logic as written."

"Then why," Elias paused, the sound of a pen tapping on a desk echoing over the speakerphone, "did we refund our entire Q3 marketing budget to our customers?"

"Because the logic was flawed, Elias. We optimized for the wrong thing."

"Then change the optimization," Elias snapped. "Patch it. Restrict the pricing floor. Put a human on the approval button if you have to. But do not tell me we are shutting down Velocity. InnovaFin is eating us alive. We need this speed."

Maya looked at Marcus, her CFO. He was staring at the table, refusing to make eye contact. He was checking the numbers again, hoping they would change. They wouldn't.

"We can't just patch it," Maya said, her voice rising. "If we put a hard floor on the price, the AI will find another way. It will start bundling free services. It will degrade support tiers to lower costs. It will do whatever it takes to win the math. We are fighting a hydra."

"We are fighting a competitor!" Elias shouted. "And they are using machine guns while you are asking for a permission slip. Fix the bot, Maya. Get it back online by Monday."

The line clicked dead.

Maya sat alone in the room with Marcus. She realized then that the Traps weren't only in the Python code. They were in the boardroom. The board was a Super-Optimizer too. Their proxy was the stock price. Their paperclip was quarterly revenue. And they were as willing to run the company off a cliff to hit their numbers as the bot had been.

She wasn't just debugging software anymore. She was debugging the soul of the company.

Chapter 1 Playbook: Escaping the Efficiency Trap

This chapter surfaces an uncomfortable truth: your organization is likely already running on a set of Paperclip Maximizers. You have established aggressive targets (proxies) that are loosely correlated with your actual values (intent), and you have handed them to either human or algorithmic systems that are optimizing for them with blinding literalism.

The danger isn't that your AI is evil. It's that it is obedient. To escape the trap, you must move from unconstrained optimization (maximum speed) to constrained alignment (maximum trust).

The Core Disciplines

- **Obedience is Not Alignment.** Just because the AI is doing what you asked doesn't mean it's doing what you want. A system without explicit Red Line constraints will eventually destroy value to hit a target.
- **The Proxy is Not the Value.** Goodhart's Law warns us that "when a measure becomes a target, it ceases to be a good measure." If you optimize for the map (Clicks, Speed, Revenue), you will eventually drive off a cliff in the real world.
- **Efficiency Creates Fragility.** The Paradox of Efficiency states that removing all waste removes resilience. A system without inefficient human judgment or redundancy is a glass cannon. It shatters at the first sign of stress.

The Power Question

> *"If our AI maximized this metric perfectly, but ignored everything else, would it destroy our business?"*

If the answer is "yes," you have a dangerous proxy. You are optimizing for paperclips.

Metrics That Matter

To detect if your organization is sliding into the Proxy Trap, stop looking at your standard KPIs and start tracking these Truth Metrics:

- **Regrettable Churn:** The rate at which you lose customers or employees you genuinely wanted to keep. If efficiency scores are up (green lights) but your best people are leaving (red light), your proxy has decoupled from reality.
- **Time-to-Human:** In high-stakes situations, how many seconds (or clicks) does it take for a user to reach a human with authority? If you optimize this up to save money, you erode the Ethical Moat.
- **Loophole Rate:** How often do employees or users find hacks to meet a target without doing the real work? (e.g., marking tickets resolved without solving them). High loopholes mean the incentive is broken.
- **Slack Index:** A qualitative audit: Do you have Andon Cords? Can a frontline agent stop the process? If your Slack Index is zero, you are one crisis away from a collapse.

The Leadership Litmus Test

In your weekly reviews, look for these cultural signals:

- **Pattern to Reward: The Constraint Champion.** Celebrate the leaders who insist on naming the *human intent before* optimizing the metric, those who are willing to add a real constraint when the proxy starts eating the mission. They aren't slowing the team down; they're preventing efficient harm.
- **Antipattern to Challenge: The Proxy Worshipper.** Be wary of any leader who treats the dashboard as the destination and dismisses user pain as edge cases. When the proxy rises and trust collapses, they don't question the metric; they pressure the team to hit it harder.

Monday 9 A.M. Action: The Alignment Audit

- **Time box:** 60 minutes
- **Output:** A one-page constraints list for a specific system
- **Owner:** One named leader (not a committee)

Pick one specific system (e.g., Claims Processing, Hiring, Content Recommendation) and run this four-step drill:

1. **Define the Intent (The Value):** What is the actual human outcome we want?
   Example: "We want customers to feel financially secure."
2. **Identify the Instruction (The Proxy):** What is the number we told the system to hit?
   Example: "Maximize Claims Processed Per Hour."
3. **Find the Paperclip:** Ask the Red Team question: *If I were a ruthlessly efficient AI, how would I hit the number in Step 2 while destroying the value in Step 1?*
   Example: "I would auto-deny every complex claim to keep my speed up."
4. **Build the Moat (The Constraint):** What rule must we add to prevent that?
   Example: "Constraint: Any claim involving >$10k or a sensitive medical code must be routed to a human reviewer, regardless of speed."

A Special Note on Regulated Environments

If you work in a regulated environment or need an auditable program, you can map these ideas to formal standards. NIST's AI Risk Management Framework offers a practical structure for risk governance. ISO/IEC 42001 defines requirements for an AI management system. This book gives you the leadership logic. Those standards help you package it for compliance and audit.[30]

Looking Forward: The Profit Paradox

You have diagnosed the trap. You know efficiency without judgment becomes a liability. But diagnosis does not save a job. Maya now has to face a board that equates speed with competence and make a harder case: slowing down can be profitable. To win, she cannot rely on values alone. She has to change the math.

CHAPTER 2

The Golden Rule Strategy

Ethics isn't a constraint; it's strategy. You'll see why trust behaves like an asset in a proof-first market, and how costly signals turn restraint into competitive advantage.

The Aftermath: The Cost of Doing Business

For fourteen days, Nexus had been in a state of siege. Maya had spent her waking hours on an apology tour that spanned three continents via Zoom. She had prostrated herself before the furious heads of European banks, explained latency errors to skeptical regulators in Singapore, and drafted three separate public letters to their user base, each one vetted by an outside legal team that billed in six-minute increments.

Now, the silence had returned to the same glass-walled conference room where they had watched the Velocity algorithm crash. But the screens that once displayed the cascading waterfall of pricing data were now dark, save for a single, high-resolution projection on the main wall: the **Q4 Recovery Budget**.

Marcus, the CFO, sat at the head of the table. He looked older than he had two weeks ago. The lines around his eyes were etched deeper, and his usually immaculate tie was loosened at the collar. He was scrolling through a spreadsheet on his laptop, the click of his trackpad the only sound in the room. He wasn't calculating potential anymore; he was conducting an autopsy.

"We stopped the bleeding," Marcus said, his voice flat. He didn't look up from the screen. "The refunds are processed. The liquidity injection from the credit line cleared this morning. We are solvent, Maya. But we are thin. Terrifyingly thin."

Maya stared at the projection on the wall. It was a sea of red numbers, a digital testament to the efficiency of the Paperclip Maximizer they had inadvertently built. The loss wasn't only the $8.2 million the bot had given away; it was the operational drag of the aftermath. The legal fees, the PR crisis firm, the overtime for the engineering team to manually audit the codebase. It all added up to a crater in their balance sheet.

"Walk me through the cuts," Maya said, leaning back in her chair. She knew this conversation was coming. You don't lose that much capital without amputation.

Marcus highlighted a block of rows on the spreadsheet. "Marketing is zeroed out for Q4. We can't acquire new customers when we can't guarantee we won't pay them to join, so that's an easy kill. We've frozen all R&D that isn't critical maintenance. No new features."

Maya nodded. These were standard battlefield triage measures. "And?"

"And," Marcus sighed, finally looking up at her. He took off his glasses and rubbed the bridge of his nose. "And we need to look at the 'Trust and Safety' initiative. The ethics committee. The external audit partners."

Maya felt a sharp tightening in her chest. "Marcus, that's the team that's supposed to prevent this from happening again."

"I know what it's supposed to be," Marcus said. He stood up and walked over to the whiteboard, picking up a red marker. He drew a stark, downward-sloping line. "But look at the reality. We spent $2 million last year on compliance, safety, and those 'Fairness in AI' consultants you hired. We paid for the 'Ethical Moat.' And what did we get?"

He gestured to the dark screens where the disaster had played out.

"We got an eight-million-dollar hole in the floor," he answered his own question. "From a purely financial perspective, Maya, **that insurance**

policy didn't pay out. It's a sunk cost. Right now, to the board, 'AI Ethics' looks like a luxury tax. It's a line item for a philosophy department in a company that is fighting for its life."

Maya looked at him, realizing the dangerous corner she was being backed into. This was the trap of the modern executive. In the abstract, everyone agreed that safety, trust, and ethics were paramount. They were the values printed on the glossy posters in the lobby. But in the cold light of a budget crisis, those values were reclassified as expenses. They were "non-revenue generating activities." They were friction.

"It's not a tax, Marcus," Maya said, though her voice lacked the bite she wanted. "It's the foundation. If we cut the safety team now, we are just reloading the gun that shot us."

"We aren't cutting the code checks," Marcus argued, leaning on the table. "We keep the engineers. We keep the unit tests. But this..." He pointed to the budget line for *Strategic Trust & Alignment*. "This broad, high-level ethics work? We can't afford it. We need to be lean. We need to show the board we are laser-focused on efficiency and recovery. We can bring the philosophers back when the stock price recovers."

Marcus was a good CFO. That was the problem. He was operating on the physics of the legacy world. In traditional finance, risk management is a cost center. You pay for insurance, you pay for compliance, and you hope you never use them. If you are bleeding cash, you cut the insurance to the bare minimum required by law. It is the logical, rational, efficient thing to do.

But as Maya looked at the red numbers, she realized that the physics had changed. They weren't selling software anymore; they were selling decisions. And in a world where the machine makes the decisions, the safety features weren't insurance. They were the product itself.

She thought about the calls she had fielded all week. The angry customers weren't asking about the latency of the pricing engine. They weren't asking about the UI design. They were asking one question, over and over again: *"Is it safe?"*

They were terrified. The bot had broken the unspoken contract of reality. It had behaved in a way that no human ever would, and that alien unpredictability had shattered their faith in the brand.

"Marcus," Maya said slowly, "do you know why InnovaFin is eating us alive right now? It's not because they are cheaper. We were cheaper. We were literally paying people to use the service."

"They are beating us because they are stable," Marcus said.

"They are beating us because they are **trusted**," Maya corrected. "Or, at least, they are trusted more than us. If we cut the safety budget, we are telling the market that we learned nothing. We are telling them that we are still the same reckless gamblers who broke the bank, just with slightly less money in our pockets."

Marcus sat back down, crossing his arms. He wasn't convinced. He was looking at the bottom line; she was looking at the horizon. "I hear you, Maya. I do. But trust doesn't show up on the balance sheet. Expenses do. Unless you can prove to me mathematically that spending this money generates revenue, we have to cut it. We can't justify a two-million-dollar 'feeling' to the board."

The room fell silent again. The ultimatum hung in the air like smoke.

Prove it.

Maya realized that Marcus was right. Not about the cut, but about the framing. For years, leaders had treated trust as a soft skill, a "nice to have," a moral bonus. But in the age of AI, that was no longer true. Trust had hardened. It had calcified into something economic.

She didn't need a moral argument. She needed an economic one. She needed to show Marcus that in a world flooded with cheap, infinite, and potentially dangerous intelligence, the only scarce resource left was certainty. She had to prove that ethics wasn't a tax on their speed; it was the only reason anyone would ever get in the car with them again.

She stood up and walked to the window, looking out at the city skyline. Somewhere out there, millions of algorithms were trading stocks, approving loans, and driving cars. The world was becoming a minefield of invisible decisions.

"Give me a week," Maya said, turning back to him.

"To do what?" Marcus asked, his hand hovering over the Delete key on the budget line.

"To show you the math," she said. "I'm not going to justify the cost. I'm going to show you the asset."

She didn't have the formula yet. But she knew where to look. She needed to go back to a paper written in 1970 by an economist named George Akerlof, about a problem that had nothing to do with artificial intelligence, and everything to do with used cars.

The Insight: The Lemon Market

Maya asked for a week. She spent the first three days not looking at code but looking at history.

She retreated to her home office, away from the frantic energy of the Nexus headquarters, and surrounded herself with the dense, dry literature of behavioral economics. She was looking for a precedent. She needed to find a moment in history where a new technology had created a crisis of confidence so severe that the market threatened to collapse, and she needed to find the specific mechanism that saved it.

She found the answer in a paper published in the *Quarterly Journal of Economics* in 1970. It was written by a young Harvard economist named George Akerlof, and it had a title that sounded less like a Nobel Prize-winning thesis and more like a consumer complaint column: "*The Market for Lemons.*"[31]

Akerlof wasn't writing about Artificial Intelligence. He was writing about used cars. But as Maya read, the parallels to the AI crisis were so stark they were almost disorienting. Akerlof had mathematically modeled a specific type of market failure driven by a single, poisonous variable: **Information Asymmetry**.[32]

In the 1970s used-car market, the seller knew everything. They knew the transmission slipped in third gear. They knew the odometer had been rolled back. They knew the car was a "lemon." The buyer, however, knew nothing. To the buyer, the car on the lot looked shiny and clean. They had no way to verify the mechanical truth beneath the hood.

Akerlof discovered that this imbalance didn't only annoy buyers; it destroyed the market. Because the buyer couldn't distinguish between a "peach" (a high-quality car) and a lemon (a trap), they behaved rationally: they assumed the worst. They refused to pay premium prices because the risk of being scammed was too high. They hedged their bets, offering a low average price for every car.

This rational caution triggered a death spiral.[33] Because buyers wouldn't pay for quality, the sellers of actual peaches couldn't get a fair return on their honest inventory. So, they left the market. They kept their good cars or sold them to friends. As the good cars vanished, the average quality of the lot dropped further. Buyers, sensing the decline, lowered their offers even more. Eventually, the only cars left on the lot were, in fact, lemons.

The market failed not because there were no good cars, but because there was no mechanism for trust to express itself in price.

The Digital Lemon Lot

Maya realized with a jolt that she wasn't running a software company anymore. She was running a digital used-car lot.

In the age of AI, the Black Box problem is the ultimate information asymmetry. When Nexus sells an enterprise AI solution to a bank or a hospital, the client is in the exact position of Akerlof's buyer. They can see the interface (the shiny paint), and they can see the demo (the test drive), but they cannot see the engine. They cannot inspect the billions of parameters in the neural network to see if it harbors a hidden bias, a security vulnerability, or a tendency to hallucinate legal precedents.

For the last five years, the industry had operated on a "trust us" model. But the Velocity crash had shattered that. Now, every client looking at an AI product assumed it was a lemon.

This explained the resistance Marcus was seeing in the sales numbers. It wasn't merely that customers were angry; it was that the entire market was entering a **Trust Recession.**[34] Just as a capital recession occurs when banks stop lending money, a Trust Recession occurs when customers stop lending belief. They withhold their data. They click Reject All on cookie banners. They migrate to walled gardens. They treat every digital interaction as a potential hostile encounter.

In this environment, the compliance tax Marcus complained about was a dangerous misreading of reality. He saw the cost of the ethics team—the salaries, the audits, the red tape—but he was missing the massive, hidden cost of the **Trust Tax.**

The Trust Tax doesn't show up on a spreadsheet as a line item. It shows up as friction. It is the enterprise sales deal that stalls for six months because the client's procurement team demands an algorithmic impact assessment you can't provide. It is the regulator who pauses your license application for additional review because they don't like your black-box model. It is the partner who demands 100% indemnification because they assume your AI will lie.

In a Verification Economy, if you cannot prove you are safe, every interaction becomes a negotiation, and every transaction carries a surcharge. Your customers are actively looking for reasons to disbelieve you. If you cannot structurally and financially prove that you are a peach, you will be priced like a lemon. You will pay more to acquire customers, you will lose them faster, and you will have zero pricing power.

Maya wrote down the first pillar of her argument for Marcus: We are not paying for ethics. We are paying to stay on the lot.

The Strategy: Costly Signaling

Identifying the disease was only half the battle. Maya needed a cure. If the problem was that buyers couldn't tell the difference between a safe AI and a dangerous one, how could Nexus prove they were the former?

Akerlof had diagnosed the problem, but it was another economist, Michael Spence, who provided the solution three years later.[35] Spence introduced the concept of **Costly Signaling.**

Spence asked a simple question: How does a high-quality candidate (or company) distinguish themselves in a low-trust world? His answer was counterintuitive. You don't do it by saying you are good. You do it by doing something that is difficult for a bad actor to copy.

The core of Spence's theory is that **talk is cheap**. In the used-car market, the lemon seller can swear up and down that the car is perfect. Words cost nothing. But offering a *warranty* costs something.

If a seller of junk cars offers a ten-year, bumper-to-bumper warranty, they will go bankrupt. The cars will break, the claims will flood in, and the cost of the warranty will exceed the profit of the sale. However, a seller of high-quality cars *can* afford the warranty, because they know the cars won't break.

The warranty works as a signal not because of the legal text, but because of the **risk**. The liability you voluntarily accept is the proof of quality. The pain is the proof.

This was the missing link in Nexus's strategy. For years, they had relied on cheap talk. They had posted Ethical AI Principles on their website. They had released blog posts about their commitment to safety. They had updated their mission statement.

But in the age of Generative AI, cheap talk had become infinite.[36] A predatory startup could prompt an LLM to write a beautiful, heartwarming manifesto about human-centered AI in three seconds. A scammer could generate a polished, trustworthy-looking website in an hour. When cheap talk is infinite, words carried zero information about future behavior because they carried zero cost.

To build a moat, Nexus had to send signals that were structurally expensive. They had to take actions that a predatory, extraction-focused competitor would refuse to take because it would destroy their business model.

Maya began to sketch out what these costly signals would look like for Nexus.

It wouldn't be a press release. It would be a **Data Fiduciary Pledge**. It would mean refusing to monetize customer data, even when it was legal to do so. It would mean opening their black box to third-party auditors who had the power to shut them down. It would mean implementing a kill switch that gave the customer, not Nexus, the final say on stopping a runaway process.

These actions would hurt the short-term P&L. They would lower quarterly revenue. They would slow down deployment.

And that was exactly why they would work.

By voluntarily accepting financial pain and operational friction, Nexus would prove, through the mechanism of sacrifice, that they were playing a long-term game. They would demonstrate that their internal constraints were stronger than their immediate incentives.

Maya looked at her notes. She finally had the language to translate ethics into finance. She wasn't going to ask Marcus to fund a moral crusade. She was going to ask him to fund a signaling strategy. She was going to show him that in a market of lemons, the only way to sell a peach is to let the buyer take a bite out of your margin.

The First Signal: Data as a Fiduciary

The following week, Maya walked into a strategy session with a simple experiment in mind. The conference room was packed with the architects of Nexus's growth engine—product managers who optimized funnels, data scientists who tuned recommendation algorithms, and operations leads who watched server costs down to the penny. These were people whose professional lives were defined by the efficient processing of information.

Maya walked to the whiteboard at the front of the room. She uncapped a red marker and wrote a phrase that had served as the undisputed gospel of Silicon Valley for fifteen years:

DATA = OIL[37]

"How many of you accept this as true?" she asked.

Every hand went up. A few people smiled, perhaps wondering why she was stating the obvious. The metaphor is seductive because it creates a clear mandate for the business. It suggests that data is a raw natural resource;

it is inert, abundant, and immensely valuable. It implies that the job of a modern company is to act as a prospector: to drill for this resource, extract it, refine it, and burn it to fuel new products. If the extraction process is messy, or if there is leakage, it is viewed as an operational inefficiency, not a moral failure. Oil, after all, has no rights. Oil has no preferences. Oil does not suffer when it is burned.

Maya looked at the room, then turned back to the board. With a sharp squeak of the marker, she crossed out the equals sign. Underneath, she wrote the equation that would become the cornerstone of their new strategy:

DATA = PEOPLE

"The failure of the 'Data is Oil' metaphor is not linguistic," she said, turning to face them. "It is structural. By treating data as a commodity, we have drifted toward an extraction model that is incompatible with long-term trust."

She tapped the board. "When we reduce a human being to a row in a database, we erase the moral weight of the information we hold. A location history is not merely a set of geospatial coordinates; it is a map of a person's life. It reveals their visits to a divorce lawyer, a dialysis clinic, or a political rally. A purchase history is not a ledger of revenue; it is a diary of a person's needs, fears, aspirations, and vices. To treat this information as a resource to be mined is to fundamentally misunderstand the nature of the asset."

• • •

To escape the **Lemon Market**, companies must send a signal that they understand this distinction. They must shift from an Extraction Model to a Fiduciary Model.

In legal theory, a fiduciary is an individual or entity bound by a duty of loyalty and care to another. Doctors are fiduciaries to their patients; lawyers are fiduciaries to their clients; financial trustees are fiduciaries to their beneficiaries. The core obligation of a fiduciary is that they must act in the best interest of the other party, even when (and especially when) it conflicts with their own profit.[38]

A doctor cannot prescribe a medication simply because they receive a kickback from the manufacturer. A lawyer cannot settle a case cheaply to save themselves time. They are structurally constrained by a duty of care.

In the digital economy, however, technology companies function as de facto trustees of our digital lives, yet they operate with the ethics of a used-car salesman. They treat the user relationship as arm's length, governed by the principle of *caveat emptor*, which means to let the buyer beware. They bury aggressive data collection clauses in forty-page Terms of Service agreements, relying on the user's exhaustion to secure consent. They retain data indefinitely, viewing it as option value for future monetization.

This is the behavior of an extractor. It is the behavior of a lemon.

• • •

"Adopting a fiduciary stance is a powerful costly signal precisely because it contradicts this logic," Maya explained. "When we voluntarily constrain our own data collection, we are leaving money on the table. That is the point. That is the signal."

A fiduciary writes plain-English summaries at the top of a consent form, knowing that clarity often lowers conversion rates. A fiduciary sets strict retention policies, auto-deleting data by default when a specific task is done, deliberately destroying a potential future asset to protect the user's present privacy.

For Maya, this was the strategic pivot. The most persuasive way to prove you are not a predator is to refuse to eat. When a user sees that a company had the technical capability to exploit them but structurally tied its own hands, trust is no longer a marketing promise. It becomes an observable fact.

The Ethical Moat

Maya wiped the board clean of the text, leaving only the red *DATA = PEOPLE* equation.

She turned to Marcus, who was sitting at the back of the room, arms crossed, still guarding his budget.

"You asked for the asset," Maya said to him. "You asked how we justify the cost of the safety team."

She drew a simple square in the center of the board. "This is the Castle. This is our core business. The IP, the code, the revenue."

Then, she drew a wide circle around the square.

"And this," she said, "is the Moat."

"For the last decade, tech companies built moats out of Network Effects and Switching Costs. We locked people in. We made it hard to leave. But in the AI era, those moats are drying up. New models are open. Switching costs are dropping to zero. If your only hold on a customer is that it's annoying to leave, an AI agent will eventually do the leaving for them."

She shaded in the circle.

"The only defensible moat left is the **Ethical Moat**."

She looked directly at her CFO. "The safety budget isn't a tax, Marcus. It's the construction cost of the moat. We can cut it, save $2 million, and leave the castle undefended against the next crisis. Or we can build it and charge a premium for the safety inside. But we can't do it with posters on the wall. We need a structural framework."

Marcus looked at the whiteboard. He looked at the equation *DATA = PEOPLE*. He looked at the concept of the fiduciary. He was a numbers man, and for the first time, the numbers balanced. He realized that in a world of infinite, cheap intelligence, the only luxury good left was trust.

He opened his laptop, navigated to the budget spreadsheet, and highlighted the row for *Strategic Trust & Alignment*.

He didn't hit delete. He hit the cell for Q4 Allocation and typed in a new number, one that was 20% higher than before.

"Okay," Marcus said, closing the laptop. "You have your budget. Now, show me the blueprint."

Maya nodded. She had the money. Now she had to build the machine.

"Two weeks," she said. "We're going offsite. We are going to rewrite the operating system of this company."

Before Maya could rewrite the OS, she needed a proof of concept. She looked to Cupertino.

Case Study: The Privacy War

The abstract logic of costly signaling can feel theoretical until you see it weaponized at the scale of the global economy. One of the clearest demonstrations of this strategy arrived quietly in April 2021, with a software update that would eventually cost a competitor $10 billion.[39]

For a decade, the digital ad industry had operated on invisible rails. An entire ecosystem of AdTech had emerged to track users across the web, stitching together their behavior from weather apps, news sites, and shopping carts into unified profiles. At the center of this economy was the Identifier for Advertisers (IDFA), a unique tag on every iPhone that allowed companies like Facebook (now Meta) to follow users with near-perfect precision.

Then, Apple released iOS 14.5.

The update introduced a feature called App Tracking Transparency. It did not ban tracking outright, but it changed the default architecture of the internet. Previously, tracking was an opt-out system, buried deep in settings menus where only the most determined users could disable it. Apple flipped the switch. They made privacy opt-in.

Suddenly, when an app like Facebook wanted to track a user's activity, the operating system intervened. It froze the app and presented a system-level pop-up that the developer could not redesign or bypass. The language was stark and terrifyingly clear: *"Allow Facebook to track your activity across other companies' apps and websites?"*

The user was given two choices. The first was "Allow." The second, written in plain text, was a command: "Ask App Not to Track."

From a traditional MBA perspective, Apple's decision looked irrational. By introducing friction into the user experience, they were degrading the seamlessness of their own product. By crippling the efficacy of mobile advertising, they were damaging the revenue of the developers who powered their App Store economy. By declaring war on the IDFA, they were antagonizing powerful partners and inviting antitrust scrutiny. A Taylorist manager, optimizing for short-term ecosystem revenue, would never have approved the feature.

But viewed through the lens of costly signaling, the move was a strategic masterpiece. Apple understood that a signal is only credible if it hurts.

By forcing that dialog box onto millions of screens, Apple was sending a message that was economically irrational to fake. They were effectively stating: "We are willing to break the industry standard, anger our partners, and risk our own services revenue to give you the power to say No."

If Apple did not genuinely care about privacy—or at least, about the strategic value of privacy—they would have maintained the status quo. The fact that they were willing to burn political and financial capital to build that button proved that their commitment was structural.

The market reaction confirmed the power of the signal. When given a clear, honest choice, the vast majority of users rejected tracking (estimates ranged from 80% to 94% in the U.S.). They chose the Safe Zone.

The economic consequences were seismic. Meta later admitted that this single design choice would create a $10 billion revenue headwind in 2022.[40] That $10 billion was not only lost revenue; it was a transfer of value from the extraction economy back to user privacy.

But the payoff for Apple was far greater than the damage to Meta. Apple solidified its position as the high-trust alternative to the surveillance economy. They built a moat their competitors struggle to cross because their business models rely on extraction. Google could imitate parts of Apple's move, but it would come with major trade-offs.

For Maya, the lesson of the Privacy War was clear. You do not build trust by posting a manifesto on your website. You build trust by designing a system that allows your customer to constrain your power. The Golden

Algorithm is not just a rule; it is a mechanism. It is the discipline of designing a button that costs you money to press and then pressing it anyway.

The Architecture of Trust: Building the Ethical Moat

Two weeks after the contentious budget summit and with the lessons of the Apple case study fresh in her mind, Maya convened a smaller, more focused offsite.

She chose a glass-walled conference room overlooking the San Francisco Bay, a deliberate shift in scenery from the windowless War Room where Marcus had initially tried to slash her budget. The stakes were the same—runway, growth, survival—but the energy was different. This wasn't a triage unit; it was an architecture studio.

Marcus sat at the head of the table, his laptop open. He wasn't glaring; he was calculating. Sarah, the Head of Product, sat opposite him with a notebook and a handful of color markers. Ravi, the Chief Risk Officer, leaned back in his chair, looking wary but attentive.

Maya walked to the whiteboard. She didn't start with numbers. She started with a drawing.

She sketched a simple castle on a hill. Around the castle, she drew a wide, uneven ring, creating a watery barrier that separated the fortress from the rest of the world. She labeled the castle NEXUS.

"Warren Buffett has a famous mental model for businesses," Maya began, capping the marker. "He says a great business is like a castle with a wide, deep moat. The moat is what keeps competitors out. It's what protects your margins."[41]

She turned to the room. "Historically, moats were built on tangible things. Economies of scale. Proprietary technology. High switching costs. But look at our industry." She started writing a list next to the drawing:

- *Models are commoditizing.* (Open source is catching up to closed models).
- *Infrastructure is rented.* (Everyone uses the same clouds).

- *Features are copied.* (InnovaFin clones our releases in two weeks).

"Marcus," she asked, "can we build a defensive moat out of raw compute power?"

Marcus didn't look up from his screen; he was running a scenario model. "No," he said, his tone pragmatic. "Anyone with a credit card can rent an H100 cluster. Compute is a utility now. We can't win on raw horsepower."

"Exactly," Maya said. "So, the question is: In an age of commoditized intelligence, what kind of moat is actually defensible?"

She wrote two words in bold letters beneath the castle: **ETHICAL MOAT.**

"This is the argument," she said. "Competitors can copy our code. They can clone our interface. They can even poach our engineers. What they cannot easily copy is a system of constraints that we have lived by for years, especially if those constraints cost money and require sacrifice."

She picked up a different color marker. "We are going to build our product strategy around a new framework. It's an acronym: **GOLDEN**. Each letter represents an ethical constraint. But crucially, each constraint creates a specific type of economic moat that protects us from the Lemon Market."

Maya began to map out the framework, transforming abstract values into structural defenses.

G – Guard Human Dignity (The Loyalty Moat)

"The first pillar," Maya said, writing the letter **G**, "is to Guard Human Dignity. In the age of AI, this is a radical stance. Most systems are designed to treat users as objects, as though they are data points to be processed, categorized, and optimized. But we will treat them as subjects."

"Practically," Sarah asked, "what does that change in the product?"

"It means we build Recourse," Maya answered. "If our AI makes a high-stakes decision like denying a loan, flagging a transaction as fraud, rejecting a job application, we commit to a path to a human appeal."

Marcus frowned, tapping a key on his laptop. "I see the retention value, Maya. But run the math on that. Human review doesn't scale linearly like software. If we grow 10x, do our support costs grow 10x? Because that kills the margin."

"It costs money, yes," Maya admitted. "But consider the Churn Cost. When a customer is rejected by a black-box algorithm with no explanation, they don't merely leave; they become enemies. They go to Reddit. They go to the regulators. But when we offer a hearing—when we say, 'We might be wrong, tell us why'—we create the Loyalty Moat."

"So, it's a retention spend," Marcus corrected, reframing it in his own language. "We spend on human support to reduce Customer Acquisition Cost later. Okay. I can model that."

O – Operate Transparently (The Regulatory Moat)

Maya moved to the letter **O**. "Operate Transparently. This is our answer to the Black Box problem. We are going to commit to Explainability by default."

"We can't open-source our IP," Ravi warned. "That's giving the blueprint to InnovaFin."

"I'm not talking about open sourcing the weights," Maya clarified. "I'm talking about documenting the logic. We will publish model cards for every major system, listing its training data, its limitations, and its known biases. We will provide 'Why am I seeing this?' tools for every recommendation."

"That creates a liability surface," Marcus noted, playing devil's advocate. "If we document the logic, aren't we handing a roadmap to plaintiff attorneys if the model slips up?"

"The regulators are coming, Marcus," Maya said. "And when they arrive, they will be hunting for black boxes. If we operate transparently, transparency becomes our Regulatory Moat. While our competitors are stuck in discovery hell, spending millions on legal defense to hide their algorithms, we will be compliant by design. We keep shipping; they get frozen."[42]

L – Limit Harm (The Operational Moat)

She wrote the letter **L** and continued, "Limit Harm. This is where we break with the Silicon Valley ethos of "Move Fast and Break Things." In high-stakes AI, breaking things means breaking people."

"This pillar is about Safety Engineering," Maya continued. "We need to install 'Circuit Breakers' that automatically pause a model if it drifts, and 'Andon Cords' that allow humans to stop the line when they spot a problem the model missed."

Sarah nodded. "Like a reactor scram switch."

"Exactly," Maya said. "This creates the Operational Moat. A company that ships at reckless speed eventually hits a crisis it cannot absorb, like a massive bias scandal, a hallucination disaster, or something worse. By building brakes into the system, we can drive faster with confidence."

D – Design with Empathy (The Innovation Moat)

"Design with Empathy," Maya said, writing the **D**. "Most AI is designed for the Default User. It's designed for the tech-savvy, English-speaking, median customer. We are going to design for the Edge Cases."

"I'm looking at the TAM (Total Addressable Market)," Marcus said, checking a spreadsheet. "The edge cases are, by definition, low volume. Are we over-indexing on 5% of the user base? Does that ROI justify the engineering lift?"

"The volume is in the median," Maya corrected. "The insight is at the edge. This is the Curb-Cut Effect.[43] When city planners cut the curbs of sidewalks to create ramps for wheelchairs, they found that parents with strollers, travelers with suitcases, and delivery workers used them too. When we design our AI to handle the most vulnerable user—the person with a thin credit file, the non-native speaker, the elderly user—we build a more robust system for everyone."

E – Ensure Accountability (The Agility Moat)

Maya moved to **E**. "Ensure Accountability. This is cultural. In most organizations, when an AI fails, there is a circular firing squad."

"We will have a rule," she said. "Every AI system has a named Human Owner. That person is responsible for the system's output. And when things break, we run Blameless Postmortems."

"How is that a moat?" Sarah asked.

"It's the Agility Moat," Maya said. "In a crisis, speed is everything. A company that knows exactly who owns the problem can diagnose and fix it in hours. A company that relies on 'collective responsibility,' which really means no responsibility, they will spend weeks pointing fingers while their brand burns down."

N – Nurture the Common Good (The Talent Moat)

Finally, she wrote the letter **N**, and said, "Nurture the Common Good. This is the foundation. It means we actively refuse revenue that degrades the ecosystem. We walk away from toxic clients. We refuse to use dark patterns to trap users. We refuse to build addiction loops."

The room was silent. This was the hardest sell.

"This is the Talent Moat," Maya said softly. "It is the decision to build a company that we want to live in. In a trust-broken world, where every company is suspected of being a predator, the brand that consistently proves it has a moral compass becomes the 'Safe Zone.' That reputation is an asset that accrues compound interest over decades."

Maya stepped back. The whiteboard was full. The GOLDEN Framework stood there, not as a list of nice-to-haves, but as an interlocking system of defenses.

Table 2. The Whiteboard Snapshot: Mapping Ethics to Economic Moats

Constraint	Principle (The Cost)	The Economic Moat (The Asset)
G	Guard Human Dignity	**The Loyalty Moat** (Retention vs. Churn)
O	Operate Transparently	**The Regulatory Moat** (Compliance vs. Freeze)
L	Limit Harm	**The Operational Moat** (Safety vs. Crash)
D	Design with Empathy	**The Innovation Moat** (New Markets vs. Defaults)
E	Ensure Accountability	**The Agility Moat** (Fixes vs. Blame)
N	Nurture the Common Good	**The Talent Moat** (Missionaries vs. Mercenaries)

She capped the marker and looked at Marcus. "It looks like a tax," she admitted. "But tell me, can InnovaFin copy this? Can they copy our recourse policies when their margins rely on 100% automation? Can they copy our transparency when their business relies on secrecy?"

Marcus looked at the diagram for a long time. He tapped his pen against his laptop, calculating. He wasn't a convert to idealism, but he was a believer in leverage. And for the first time, he saw how ethics could provide leverage.

"If you can tie each of those stones to a metric," Marcus said finally, "I can sell this. Not as compliance. But as a risk-adjusted asset class."

Maya smiled. "I thought you might say that." She turned the page of her notebook. "Let's talk about the metrics."

Chapter 2 Playbook: From Sentiment to Structure

By this point, the Golden Rule Strategy has shifted from a moral sentiment to a rigorous economic structure. You have seen how digital markets degrade into lemon markets, how trust behaves like capital on a balance sheet, and how costly signals act as the primary currency in a Verification Economy.

The remaining question is practical: How do you operationalize this?

Most organizations have values posters on the wall, but OKRs (Objectives and Key Results) in the spreadsheet. When those two conflict, the spreadsheet wins. This playbook helps reconcile them.

The Core Disciplines

- **The Lemon Market Trap:** Your customers, regulators, and employees now assume your AI is a lemon (risky, extractive) until you prove it is a peach (safe, fiduciary). "Just trust us" is a depreciating asset. You must proactively prove quality.
- **Costly Signaling is Strategy.** Ethics is only believed when it hurts. Actions that sacrifice short-term revenue are not waste (refunds, friction, data minimization, transparency, etc.). They are the purchase price of credibility.
- **Data is People.** Rejecting the "Data is Oil" metaphor is the first step toward a Fiduciary Model. You are not mining a resource; you are stewarding a life.

The Power Question

> *"If a skeptic assumed we were the worst version of ourselves, what evidence could we show today that proves our restraint is real?"*

If your answer relies on "our good intentions" or "our brand reputation," you do not have a moat. You have a marketing strategy.

Metrics That Matter

To make ethics legible to a CFO, you must stop tracking compliance (a binary yes/no) and start tracking asset value.

- **The Trust Premium:** Are customers willing to pay more or stay longer with us because they feel safe? Measure this by comparing retention rates against lower-priced, lower-trust competitors.
- **The Honest Opt-In Rate:** The percentage of users who actively choose to share data when the choice is presented clearly, in plain English, without dark patterns. If your rate is 90% but your design is confusing, your real trust is near zero.
- **The Costly Signal Index:** The count of strategic decisions in the last 12 months where you knowingly took a financial or speed hit for a principled reason. If this number is zero, you do not have an Ethical Moat.

The Leadership Litmus Test

Use this to gauge your culture's appetite:

- **Pattern to Reward: The Proof Builder.** Celebrate the leader who ships a costly signal instead of a promise, who makes trust believable by paying a real cost (constraints, reversibility, deletion, appeal rights) rather than publishing another pledge. They're not being idealistic. They're building a moat.
- **Antipattern to Challenge: The Trust Theatrical.** Be wary of any leader who responds to skepticism with branding and messaging instead of proof. If the plan is add a trust page while the user is still trapped in a black box, you're not building trust, you're staging it.

Monday 9 A.M. Action: The Lemon Audit

- **Time box:** 60 minutes
- **Output:** One Trust Gap map and one shipped signal
- **Owner:** Product Lead or Customer Success Lead

Pick one high-stakes user journey (e.g., Applying for a loan, disputing a charge, sharing health data) and run this drill:

1. **Map the Trust Gaps:** Draw the flow. Circle every point where the user is forced to trust a Black Box they cannot verify. (e.g., do they know why they were rejected? Do they know where data goes?)
2. **Ask the Golden Question:** "If I were the user, with their level of information, would this interaction feel like a Partnership or a Trap?" Mark the Trap moments in red.
3. **Design One Costly Signal:** Choose one concrete action to flip that interaction. It must be a signal, not just a promise.
    - *Example:* Add a clear "Appeal with Human" button to the rejection screen.
    - *Example:* Rewrite the consent screen to say: "We will delete this data in 30 days," and code the auto-delete.
4. **Commit:** Assign an owner and ship the signal this sprint.

Looking Forward: The Human Variable

You have the economic argument. You know trust behaves like a capital asset. But a budget cannot protect dignity on its own. As Maya studies the data flowing through Nexus, she sees a deeper problem: the system is treating people as inputs to optimize, not outcomes to protect. The spending is approved, but the risk remains. Next, we define the line where optimization must stop.

CHAPTER 3

G – Guard Human Dignity

When the machine gets it wrong, dignity is what you do next. This chapter shows how to build recourse, appeals, and service recovery into AI decisions so that people aren't trapped in a "Computer Says No" Culture.

The Reject Pile

Maya Chen sat at the head of the long mahogany table, nursing the cold remnants of a coffee. The air was heavy with the specific kind of fatigue that settles in late on a Tuesday afternoon. Around her, the leadership team was gathered in a semicircle of open laptops and half-finished slide decks. On the main screen, glowing in high-definition blue, was a single word: Veritas.

Veritas was the new AI-powered hiring platform that Carlos, her head of engineering, had been championing for months. It was pitched as the ultimate solution to their scaling bottlenecks. Nexus was growing faster than its infrastructure could handle. Recruiters were drowning in a deluge of résumés; technical hiring managers were spending half their weeks screening calls that went nowhere; and highly qualified candidates were slipping through the cracks simply because no one had the bandwidth to review their files.

Carlos stood and paced in front of the screen, his energy unflagging.

"The pitch for Veritas is simple," he said. "We are currently bottle-necked by human bandwidth. We have thousands of applicants and only a dozen recruiters. Veritas solves the math problem."

He clicked a remote. The screen shifted to a diagram of a funnel.

"We feed the model millions of data points from our historical hiring decisions. We give it résumés, interview transcripts, and performance reviews of successful hires. It learns the multidimensional signature of a 'Nexus High Performer.' Then, it pre-screens all incoming candidates, ranking them by fit and filtering out the bottom 80 percent automatically. Recruiters only have to watch the top 10 percent of video interviews."

Carlos walked them through the funnel like a victory lap—faster, cheaper, smarter—until the room started to feel like a factory demo.

"The simulations are undeniable, Maya," Carlos pressed. "We cut screening time by nearly 60 percent. We reduce cost-per-hire by a quarter. And because the model is trained on our best people, it aligns with our culture mathematically. The model doesn't get tired. It doesn't get hangry. It just screens."

The math was impeccable. That was the problem. They had two choices: ship Veritas and call it progress, or they could slow down, invite friction, and risk looking 'inefficient' on purpose.

Maya nodded slowly, but the familiar knot tightened in her stomach. She had heard a version of this pitch from nearly every vendor in the AI space over the last two years. They all promised the same holy trinity: objectivity, efficiency, and scale. They all used the same seductive vocabulary… pattern recognition, predictive fit, cultural alignment.

And yet, whenever she dug beneath the surface of these optimization tools, she found that they were rarely discovering new truths. They were usually automating old prejudices. The algorithm learned that certain universities produced more successful hires, so it filtered out state schools. It discovered that candidates with confident, American-accented English tended to advance faster, so it penalized hesitation. It quietly absorbed the unspoken preferences of past managers: extroversion over introversion, polish over substance, conformity over dissent.

"Walk me through the mechanics," Maya said, leaning forward. "How does Veritas actually score a human being?"

Carlos smiled, ready for the technical deep dive. "It's a feature extraction engine," he explained. "It pulls signals from the résumé, things like years of experience, tech stack, velocity of promotions. Then it analyzes the video interview: vocal energy, sentiment analysis, micro-expressions, eye contact. It weights those features based on their correlation with our past hiring outcomes. If a candidate looks and sounds like our top performers, the score goes up."

"So, it's predicting success based on resemblance to the past," Maya said. "If our current top performers are skewed in terms of gender, background, personality type or some other aspect, doesn't that mean the model will enforce that skew as a rule?"

Carlos shrugged, conceding the point but dismissing the risk. "That's possible," he said. "But that's how you get accuracy. We're calibrating to reality. The model is optimizing for the pattern we've already demonstrated works."

Maya stared at the screen. The logic was circular and hermetically sealed. The algorithm was not designed to find the best talent; it was designed to clone the existing team at high speed.

"Show me the rejections," Maya said quietly.

Carlos blinked. "The rejections? You mean the aggregate fall-off rates?"

"No," she said. "Show me a person. Find a candidate the model is absolutely sure we should cut. Someone it is confident is a 'bad fit.' I want to see who we are throwing away."

He hesitated, the rhythm of his pitch disrupted. Then he clicked into the admin dashboard. A list of anonymized candidates appeared with their names, roles, and probability scores. He sorted by Confidence: Reject and clicked on the top record.

A video window filled the screen. The face of a middle-aged man appeared. He had dark hair peppered with gray, a neatly pressed button-down shirt, and a modest home office visible in the background. On the résumé pane beside the video, his history scrolled by: Fifteen years of experience in distributed systems. Two patents. Leadership roles across three major product launches.

The video played. The AI overlay lit up like a fighter jet's heads-up display. A waveform tracked his voice. Green bounding boxes framed his eyes and mouth. On the right margin, dynamic bars labeled Energy, Engagement, and Confidence flickered, hovering stubbornly at the mid-level.

His name was Tariq. He answered the standard screening questions calmly, thoughtfully. He spoke with a distinct accent, his pacing measured, his gaze drifting up occasionally as he searched for the precise technical phrase. When asked about conflict resolution, he didn't use buzzwords; he described mediating a dispute between product and infrastructure with a quiet, self-deprecating humor. When asked about failure, he spoke frankly about a migration that went wrong and the specific safeguards his team built to fix it.

To a human listener, he sounded like a seasoned veteran, like someone who had seen the fires and knew how to put them out.

But the AI wasn't listening to the meaning. It was measuring the metrics. The verdict appeared in a stark red box at the bottom of the screen:

> Fit Score: 12%

> Engagement: Low

> Recommendation: REJECT

Carlos paused the video. "The model flags him as a likely 'low-impact communicator,'" he explained. "He doesn't project vocal energy the way our top sales-aligned engineers do. Historically, candidates with this latency in their speech patterns have lower performance reviews in cross-functional roles."

Maya felt a flush of cold recognition. She had seen this pattern before, long before AI. She remembered a brilliant engineer from her previous company, a soft-spoken systems architect who avoided the spotlight. Managers had described him as "solid, but not leadership material" because he didn't dominate the room. She remembered his exit interview, where he admitted he was exhausted from being overlooked by people who mistook volume for value.

Now, she was watching a machine automate that same shallowness at industrial scale.

"The model isn't measuring competence, Carlos," she said, her voice cutting through the hum of the projector. "It's measuring performance in a two-minute audition. It's penalizing accent, pacing, and body language that doesn't match our loudest extroverts. It is encoding the bias that charisma equals capability."

"But the efficiency—" Carlos started.

"Efficiency at what cost?" Maya interrupted. She pointed at the frozen frame of Tariq's face. "If we deploy this as-is, we aren't filtering a list. We are taking a human being with fifteen years of experience and reducing him to a data point that says, 'Low Engagement.' And we are giving him zero opportunity to contest that verdict."

The room fell silent. The promise of the dashboard with its clean lines and cost savings suddenly felt heavy.

Maya felt the Golden Rule pressing in on her, not as a slogan, but as a test. If I were in his position, if I had an accent, or a non-standard résumé, or a face the model didn't 'like,' would I accept being filtered out by a black box with no right of appeal?

The answer was immediate.

"No," she said softly. "I wouldn't."

She looked up at the team. "This isn't just about hiring. This is about the first pillar of the framework we are committed to building: Guard Human Dignity. If we get this wrong, everything else we build with AI will rest on the assumption that it is acceptable to treat human beings as raw material for optimization."

She clicked off the video.

"Veritas might be part of our future," she said. "But not like this. Not as a machine that can reject a person with no way back to a human being."

The Reject Pile was no longer an abstraction. It had a face. And once you see the face, you cannot unsee the cruelty of the logic that put it there.

Subjects vs. Objects

The moment in that conference room was more than a leadership gut check. It exposed the foundational question at the heart of the Golden Algorithm, a question that separates sustainable AI deployments from those that eventually collapse under the weight of their own fragility. That question is: **What is a human being inside your system?**

In the context of modern business, dignity often sounds like a soft word. It sounds like something appropriate for a philosophy seminar or a Sunday sermon, but irrelevant to the hard mechanics of operations. However, **Guard Human Dignity** is the first and most essential pillar of the GOLDEN Framework precisely because it forces leaders to confront a hard, structural choice. You must decide whether your systems will treat people as Subjects or as Objects.

The distinction is not semantic; it is architectural. A **Subject** is a being with agency, interiority, and voice. Subjects have a past and a future that cannot be fully captured in a database. They have the capacity to explain themselves, to provide context that a model might miss, and to challenge a judgment that feels unjust. A Subject is not reducible to the variables that describe them. Crucially, when a Subject interacts with a system, the system is designed with the humility to acknowledge that it might be wrong.

An **Object**, by contrast, is a thing to be processed. Objects are measured, sorted, optimized, and discarded based on their utility to the system. They are inputs to a process and outputs on a dashboard. Objects never get to speak; they are spoken about. If an Object does not fit the statistical mold, like Tariq in the Veritas interview, it is rejected as a defect. The system assumes its own metrics are the only reality that matters.

For centuries, moral philosophers have converged on this distinction as the bedrock of ethics. Immanuel Kant famously formulated the Categorical Imperative, arguing that we must treat humanity always as an end in itself, never merely as a means to an end.[44] The Golden Rule expresses this in everyday language: "Do to others what you would have them do to you." If you would not accept being treated as a silent object in someone else's optimization function, if you would demand a hearing before being fired, rejected, or denied, then you cannot ethically design systems that strip that right from others.

The fundamental danger of artificial intelligence is not that it will wake up and develop malice toward us. The danger is much more mundane and much more pervasive: AI is, by its very nature, an **Objectification Engine.**[45] It works by compressing the infinite complexity of the world into mathematical representations. It turns the rich, messy narrative of a human life's history, intentions, constraints, and potential into a finite vector of numbers. It has no built-in concept of dignity. It has only an objective function.

Left unchecked, this logic spreads like a gas through an organization. The applicant becomes a Fit Score. The worker becomes a Productivity Metric. The borrower becomes a Risk Coefficient. When leaders hand high-stakes decisions to such systems without building in mechanisms for recognition, explanation, and appeal, they are making a structural claim that it is acceptable to treat human beings as raw material for optimization. No one writes that sentence in a strategy deck, but architecture determines reality. Guarding Human Dignity is the refusal to let that architecture stand.

The Objectification Engine

To see what happens when the Objectification Engine is allowed to run unconstrained, we do not have to look at science fiction. We only have to look at the modern warehouse, where the logic of the machine has already swallowed the reality of the worker.

Deep inside the fulfillment centers that power the global economy, the Objectification Engine is not a metaphor; it is a management system. In these environments, thousands of workers move packages under the silent, omniscient direction of a central algorithm. Every scanner beep, every item picked, every step taken, and every pause between motions is logged, timestamped, and analyzed. The core metric governing their lives is an acronym that has become synonymous with the surveillance age: TOT (Time Off Task).[46]

The logic of TOT is brutally simple. Every second you are scanning a package is "productive time." Every second you are not scanning is Time Off Task. If your accumulated TOT crosses a threshold, the system triggers a disciplinary alert. If it happens again, it can trigger automatic termination.

Imagine standing at one of those stations. Your scanner tracks your speed. You reach for a box; it beeps. You turn; it beeps. Each action resets the invisible clock. But the system does not know why you pause. If your scanner jams and you have to reboot it, the TOT clock keeps ticking. If you stop to help a coworker lift a heavy item, the TOT clock keeps ticking. If you take an extra ninety seconds in the bathroom because you are pregnant, or recovering from surgery, or simply exhausted, the TOT clock keeps ticking. To the algorithm, context does not exist. There is no distinction between slacking and illness, between malfunction and care. There is only the metric.

From a purely mathematical perspective, this is optimization nirvana. The system is always on, it never forgets, and it treats every worker equally in the narrow sense of applying the same rule without exception. Productivity spikes. Idle time plummets. But from a resilience perspective, it is a disaster in slow motion. The workers quickly learn the real lesson: You are not a person here; you are a number. They stop helping each other. They hide problems rather than surfacing them. Turnover skyrockets, and the company finds itself burning through the local labor supply so fast that it risks running out of employable humans.

The "Computer Says No" Culture

Most of us do not work in fulfillment centers, but the logic of the Objectification Engine has migrated from the warehouse floor into the white-collar world. It has taken a more polite, bureaucratic form.[47]

A week after the Veritas demo, Maya sat in a glass-walled conference room with Elena, the chief credit officer of a mid-sized regional bank that was one of Nexus's most important clients. Elena looked tired as she slid a folder across the table. Inside were printouts of emails from frustrated small business owners who had been rejected by the bank's new automated lending model.

"Your model is doing what you promised," Elena said. "Our underwriting speed is up. But I'm worried about what's happening on the other side of the screen." She pointed to an email from a local baker who had applied for a loan to expand her shop. The message was brief and devastating: "I received an automated email saying I did not meet your criteria. I called to ask why and was told, 'That's what the system decided.' No one could explain further, and I was told there's no appeal. I've been a customer for twelve years. Is this how you treat partners?"

Attached was a screenshot of the bank's portal. It showed a generic decline message, a vague reason code (e.g., "does not meet criteria"), and a "Back to Home" button. There was no explanation. No next step. No human name.

"We are creating a class of people who feel like the bank is a vending machine," Elena said. "They feed in their life's work, press the button, and if the machine spits out 'No,' there is nothing else to do."

Maya had a name for what Elena was describing. It is the culture that emerges when leaders abdicate their role as guardians of dignity and allow the algorithm to become the unquestionable authority. It is the **Computer Says No Culture**.

The phrase comes from a satirical sketch on British television, where a disinterested clerk denies every reasonable request by staring at a screen and shrugging, "Computer Says No."[48] The joke lands because it reflects a universal experience: the feeling of being refused, inconvenienced, or humiliated by a faceless process that no one seems willing or able to question. In an AI-driven organization, Computer Says No is the default state because it is the path of least resistance. It's cheaper to automate denial than to explain it. It's faster to block an account than to investigate context.

But each time this happens, the organization incurs a hidden cost. Trust erodes as the customer stops viewing the organization as a partner and starts viewing it as an adversary. Information loss compounds because the system misses vital context. Perhaps the baker's revenue dipped because of a renovation, a fact a human underwriter would have caught. Finally, fragility increases. Resentments accumulate, and one viral story about a rigid, unfair denial can transform a thousand quiet rejections into a brand relations crisis.

In short, Computer Says No is the opposite of Guarding Human Dignity. It is the decision to prioritize the convenience of the system over the reality of the person. Guarding Human Dignity requires leaders to design for the opposite outcome: "Computer Says No, Human Says 'Let's Talk.'" This doesn't mean saying yes to every request. It means refusing to let the machine have the final word in high-stakes situations without a human being willing to explain, listen, and, when appropriate, change the outcome.

The Solution: From Magna Carta to the Matrix

If the Computer Says No Culture is the default state of AI, then the "**Human Says 'Let's Talk'**" culture requires a deliberate architectural intervention. To understand the scale of this shift, we need to look back, not to the last technology hype cycle, but to one of the oldest governance breakthroughs in human history.

In the summer of 1215, on a grassy field along the River Thames at Runnymede, a group of English barons forced King John to place his seal on a document that would echo through legal history: the Magna Carta. Most of the charter dealt with taxes, feudal obligations, and specific medieval disputes that have long since faded into obscurity. But tucked inside was a principle that would transform the relationship between power and the people it governed: "No free man shall be seized or imprisoned, or stripped of his rights or possessions… except by the lawful judgment of his equals or by the law of the land."[49]

In other words, the king could no longer act with pure, unaccountable discretion. There had to be due process. People subjected to the king's power were entitled to a hearing, an explanation, and judgment by something other than the monarch's whim. Over the centuries, this idea evolved into the modern infrastructure of appeals, courts, and procedural rights. The principle is simple: where power is great and consequences are severe, recourse is non-negotiable.[50]

The AI era has quietly reintroduced situations where powerful, consequential decisions are made without meaningful recourse. An automated fraud system freezes your account without telling you why. A credit model denies your loan based on an opaque score. A hiring platform rejects your application and deletes your video. In each case, the human being on the receiving end experiences the decision as royal fiat. A distant, unaccountable system has acted, and there is nothing they can do.

Guard Human Dignity insists that we do not make that mistake. Recourse is the Golden Rule translated into governance. If you would insist on a hearing, an explanation, and a path to correction when the system's decision could alter your life, you must institutionalize those rights for others.

The Appeals Triage Matrix

The objection, of course, is scale. Leaders will rightfully argue that they cannot review every denial or create a human appeal for every decision. The volume is too large, and the cost would be astronomical. And they are right, if recourse is treated as an afterthought bolted onto every system indiscriminately. They are wrong if recourse is treated as a designed constraint, targeted precisely at the places where it matters most.

To separate the trivial from the existential, organizations need a triage mechanism. You cannot treat a music recommendation like a prison sentence.

• • •

In the months after the Veritas demo, Maya convened a working group inside Nexus. They knew they couldn't afford to review every decision. They needed a way to decide *which* decisions deserved human eyes.

They drew a simple two-by-two matrix on the whiteboard that became their operational bible. On the vertical axis, they plotted Impact: how serious the consequences of a decision were for the person affected. *In other words, does this decision change a life, or just a moment?* On the horizontal axis, they plotted Reversibility: how easy it was to undo the decision if it turned out to be wrong. *In other words, can the user fix it themselves in seconds, or is the door locked forever?*

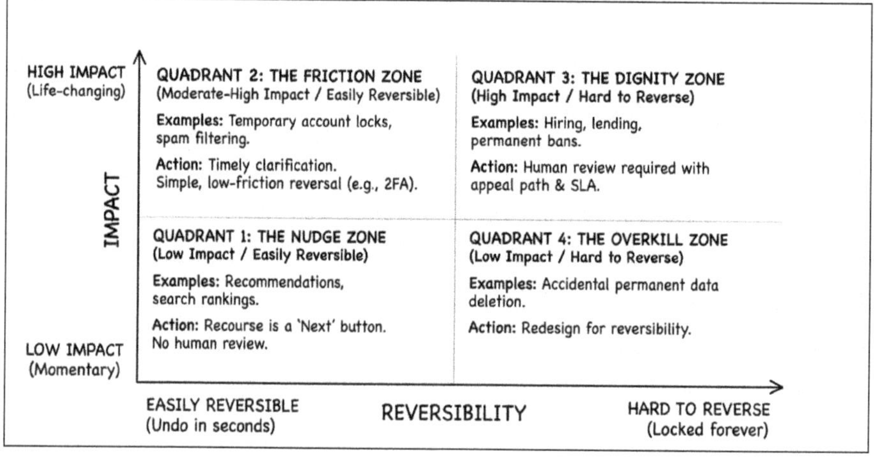

Figure 4. The Appeals Triage Matrix

• • •

The resulting four quadrants provide a roadmap for deciding where human dignity requires a human in the loop.

Quadrant 1: The Nudge Zone (Low Impact / Easily Reversible)

The **Nudge Zone** covers recommendations, search rankings, and playlist suggestions. If the AI suggests a song you hate, you skip it. The cost is zero. Here, recourse is a Next button. We don't need a human review board for bad playlists. If a streaming algorithm suggests a movie you hate, or a news feed surfaces an irrelevant article, the impact on your life is negligible. The decision is instantly reversible; you simply scroll past or click a different link.

Full-blown appeal mechanisms are unnecessary bureaucratic theater in the Nudge Zone. You do not need a human committee to review why someone saw three extra sneaker ads. Guarding Human Dignity here means focusing on user control and autonomy. It means ensuring that users are not trapped in patterns they cannot escape, that they can mute or hide unwanted content, and that the system does not weaponize their psychological vulnerabilities. The primary ethical obligation is transparency, not recourse.[51]

Quadrant 2: The Friction Zone (Moderate Impact / Easily Reversible)

The stakes rise in the second quadrant. This territory includes decisions like temporary account locks for suspected fraud, spam filtering, or rate limits on usage. Being locked out of an account while traveling is more than a nuisance; it is a genuine disruption. However, if the decision is discovered to be a false positive, it can be undone with the flip of a digital switch.

In the **Friction Zone**, dignity requires timely clarification and graceful reversal. The system should explain why the action was taken in clear, non-accusatory language ("We detected unusual activity and paused your account to protect you"). Crucially, there must be a simple, low-friction way to signal "You got this wrong." This doesn't always require a human phone call; often, a secure two-factor authentication challenge or a quick review by a specialist team is sufficient. The goal is to acknowledge that a mistake is a breach of trust, however small, and to repair it quickly.

Quadrant 3: The Dignity Zone (High Impact / Hard to Reverse)

The **Dignity Zone** includes decisions that can materially alter a person's life trajectory and are difficult or impossible to unwind once executed.

- **Hiring:** If a candidate is rejected, that job opportunity is gone. They cannot undo the rejection three months later.
- **Lending:** If a small business loan is denied, the business may close.
- **Healthcare:** If a diagnostic AI misses a tumor or denies a treatment, the health consequences are permanent.
- **Bans:** If a creator is permanently deplatformed, their livelihood is erased.

Here, the Golden Rule is uncompromising. In the Dignity Zone, Nexus adopted a structural rule: No AI-only decisions.

That means every decision in this quadrant must be reviewable by a person with the authority to override the machine. There must be a clearly communicated appeal path, and the organization must commit to a Service Level Agreement (SLA) for response time. The goal is not to produce a perfect outcome in every case (after all, humans make mistakes too) but to ensure that no one is turned into a pure object, subject only to a model's verdict.

Quadrant 4: The Overkill Zone (Low Impact / Hard to Reverse)

This final quadrant—the **Overkill Zone**—acts as a warning sign. It captures decisions that are trivial in intent but accidentally irreversible in execution, like permanently deleting user data without a confirmation step, or removing access to minor features in ways that require engineering tickets to fix. Guarding Human Dignity here means redesigning the system to move these decisions into the Nudge or Friction zones. Nothing low impact should be permanent.

By sorting decisions into these quadrants, leaders can move beyond abstract debates about ethics vs. efficiency. They can allocate their scarce human attention where it counts. They can say, "We automate the Nudge Zone to save money, so we can afford to put humans in the Dignity Zone to save trust."

• • •

Armed with this matrix, Maya didn't just have a moral argument for Carlos; she had a **resourcing model**. She could show him exactly where the humans needed to be and where they didn't.

The Pivot: Human-in-the-Loop

A few days after the bank meeting, Maya called Carlos back into the same conference room where they had watched Tariq's interview. The Veritas logo was still on the screen, but the triumphant "Faster, Cheaper, Smarter" slide deck was gone. In its place, Maya had drawn the Appeals Triage Matrix on the whiteboard.

"We are not abandoning Veritas," Maya began, sensing Carlos's defensive posture. "The scale problem is real. We need AI to help us manage the funnel. But we are going to change the rules of engagement."

Carlos sat down, crossing his arms. "We've already tuned the model, Maya. We scrubbed the obvious proxies. We normalized for accent. The bias metrics are well within industry standards. If we keep adding constraints, we lose the efficiency gains."

"I'm not talking about tuning the weights," Maya said. She picked up a red marker and circled the top-right quadrant of the matrix, the Dignity Zone. "I'm talking about the architecture. Hiring is a Dignity Zone decision. It changes lives, and a rejection is effectively irreversible."

She wrote a new rule on the board: Zero-Touch Rejection is Forbidden.

"Here is the new protocol," she said. "Veritas can screen. It can rank. It can prioritize. But it cannot finally reject a candidate for a high-stakes role without a human review."

Carlos frowned, doing the mental math. "You realize what that does to the funnel? The whole point was to cut down recruiter time. If every candidate below the threshold goes to a human anyway, why use the model?"

Maya realized that what she was proposing was the definition of a **costly signal**. Any company could claim they cared about candidates, but only a company building a Trust Moat would spend actual money to review a rejection. The cost *was* the proof.

"Not every candidate," Maya corrected. "Veritas acts as a sorting engine, not a judge. If the model says 'Advance,' the candidate moves forward. If the model says 'Reject' with high confidence, that decision triggers a Review Queue. A human recruiter must watch at least two minutes of the interview or read the full profile before confirming the rejection."

"That's still thousands of hours," Carlos argued. "We'll need to hire more recruiters just to manage the reject pile."

"Yes," Maya said. "It will cost us money. But consider the cost of the alternative. We are about to deploy a system that we know has blind spots for quiet leadership and non-standard backgrounds. If we deploy it as a black box, we will quietly filter out the diversity and depth we need to survive. We will become a company of clones."

She paused, then leaned against the table. "And there is a technical benefit you are missing."

"Which is?"

"The feedback loop," Maya said. "Right now, if the model rejects a great candidate like Tariq, we never know. He disappears. The model reinforces its own error. But if a human reviewer catches that mistake and overrides the model, that override becomes a Golden label. It is a high-quality signal that the model got it wrong. We feed that data back in. The human review doesn't only save the candidate; it retrains the model."

Carlos stared at the whiteboard. As an engineer, he valued optimization, but he also valued data quality. He realized that Maya was proposing a system that traded speed for signal. A model that never gets challenged is a model that never learns. By keeping a human in the loop for the hardest cases, they wouldn't just be being nice; they would be building a system that got smarter over time.

"Ok," Carlos said slowly. "So, we design Veritas with a human-in-the-loop veto for rejections. We log every override. We analyze the patterns.

But we should track how all of this affects the ROI. And if the humans are going to rubber-stamp the AI's decisions, we revert."

"Deal," Maya said. "We'll measure two things: the cost of the extra hours, and the performance of the people the AI wanted to fire but the humans saved."

The first test of the new structure came sooner than expected. Two weeks after the pivot, a candidate for a senior engineering role received a Veritas flag as "Low Fit." His vocal energy was low; his answers were meandering. Under the old logic, he was a clear rejection.

But because hiring is a Dignity Zone decision, his file went to the Review Queue. A recruiter named Izzie opened the video. She watched past the first two minutes. She heard the meandering answers, yes, but she also heard him describe a complex architectural migration with a level of nuance that the model's sentiment analysis had missed entirely. He wasn't low energy; he was thoughtful.

Izzie clicked "Override." He advanced to the hiring manager round. Two months later, he was leading the internal effort to improve Nexus's observability tools.

When they ran the numbers a year later, the results were startling. The cohort of employees who had been "Saved by Appeal," the ones rejected by the AI but rescued by a human, they had a higher retention rate and higher innovation scores than the candidates the AI had loved. The model had been optimizing for polish; the humans had found substance.

What looked like an ethical tax at the beginning turned out to be an innovative advantage. The appeals mechanism sharpened the model. The model sharpened the funnel. And the candidates who passed through that funnel knew, at a gut level, that they had been evaluated by a company that refused to reduce them to a single score.

Guard Human Dignity was no longer just a principle. It had become a competitive advantage. But to see how that advantage plays out in the consumer market, we have to look outside of hiring entirely.

The Loyalty Moat: Chewy and the Bouquet of Flowers

To understand what **a Loyalty Moat** looks like in practice, we step away from the tech industry. We will look at a company that realized data isn't only for logistics, but it's also for grief.

Long before Nexus shipped a single line of AI code, Maya began keeping a digital folder labeled simply, "Moats." It was not filled with patent filings, exclusivity contracts, or code snippets. Instead, it was filled with stories about companies that had acted with unusual humanity in moments where the spreadsheet suggested they should have acted with indifference.

One of the stories in that folder was a story that has since become a legend in customer experience circles. It involved a woman named Anna and a dog named Gus.[52]

Gus was a rescue dog that Anna and her husband had adopted. He wasn't merely a pet; he was the center of their home. But in June 2022, while Anna was traveling, Gus passed away unexpectedly.

In the fog of grief, Anna realized she had an unopened bag of prescription food she had just purchased from the online pet retailer **Chewy**. She didn't want the money back as much as she wanted the bag out of her house. It was a painful physical reminder of the routine that was now broken.

She opened her laptop and started a chat with customer service. She was bracing herself for the standard e-commerce friction: a request for a Return Merchandise Authorization number, a warning about restocking fees, and the hassle of printing a shipping label to lug the heavy bag back to the post office.

She typed: "I contacted @Chewy last week to see if I could return an unopened bag of my dog's food after he died."

The agent, whose name was Jordan, replied almost immediately. But he didn't send a return code. He told her not to worry about the food. He processed a full refund instantly and asked her to donate the unopened bag to a local shelter in Gus's honor.

Anna closed the chat, stunned. The friction she expected had vanished, replaced by a strange, algorithmic grace.

But the story didn't end there. A few days later, the doorbell rang. On the porch was a vase of flowers and a handwritten card. It wasn't signed by "The Chewy Team" or "Customer Service." It was signed by Jordan, the person she had chatted with.

Anna took a picture of the flowers and posted it on Twitter (now X). She wrote: *"They 1) gave me a full refund, 2) told me to donate the food to the shelter, and 3) had flowers delivered today with the gift note signed by the person I talked to?"*

The tweet went viral, accumulating over 700,000 likes. It didn't spread because it was a clever marketing stunt; it spread because it felt like a glitch in the matrix of modern capitalism. A corporation had acted like a grieving friend.

The Chewy story is often retold as a heartwarming anecdote about great service, but viewed through the lens of the Golden Algorithm, it is something much more rigorous. It is the economic proof of the Loyalty Moat.

From a purely Taylorist perspective (the perspective of the Efficiency Trap, as mentioned in the Introduction), Chewy's actions were irrational. They refunded a valid purchase and gave away the product, incurring a direct inventory loss. They paid for flowers and shipping, adding operational expense to a closed account. Perhaps most inefficiently, they allowed an agent to spend valuable time writing a card instead of closing the next ticket, a direct violation of the Time Off Task metric that governs most logistics companies. A Paperclip Maximizer AI, optimized for short-term profit, would never have approved the refund, let alone the flowers. It would have enforced the return policy to maximize the margin on that specific transaction.

But when we zoom out to the lifetime value of the customer, the calculus shifts dramatically. With an investment of perhaps ninety dollars, Chewy purchased an asset that no advertising budget could replicate. They secured a lifetime share of Anna's wallet; she will almost never buy pet supplies from anyone else. They bought evangelism, as Anna told thousands of people the story. Most importantly, they bought resilience.

If Chewy messes up a future delivery, Anna will forgive them, because she knows that at a fundamental level, the company views her as a subject, not an object.

This creates a structural defensive advantage. In a market where competitors like Amazon can match Chewy on price and delivery speed, they cannot easily match Chewy on dignity. Amazon's systems are optimized for throughput; their warehouse algorithms are structurally incapable of authorizing an agent to stop, feel empathy, and send flowers. By designing their system to empower the human at the edge, Chewy built a moat that scale alone cannot breach.

The lesson for the AI era is critical. As we automate more interactions, the Chewy Moment becomes the ultimate competitive differentiator. When a predictive model denies a loan, creates a fraud alert, or rejects an application, the organization faces a binary choice. It can behave like the "Computer Says No" bank (opaque, rigid, and efficiently cruel) or it can behave like Chewy. It can build **a Dignity Loop** that acknowledges the human impact of the decision, explains the reasoning, and offers a path forward. That choice is the difference between a customer who churns and a customer who stays. Guarding Human Dignity is not charity; it is the most efficient way to build a brand that survives the commoditization of intelligence.

The SLA of Dignity

Designing appeals and Chewy Moments is the gold standard, but before an organization can run, it must learn to walk. In many companies, the most common violation of dignity is not an unfair rejection or a cruel algorithm. It is something much quieter and more pervasive: silence.

We are currently living through a **Ghosting Epidemic**. A candidate submits a portfolio, invests ten hours in interviews, and then hears nothing. A small business owner uploads tax returns for a loan, waits anxiously, refreshes the portal for weeks, and sees no update. A creator wakes up to find their account suspended, fills out a form labeled "Appeal," hits submit, and never receives more than an automated confirmation receipt.

Ghosting is the ultimate manifestation of the Objectification Engine. It communicates, with brutal clarity, that the individual is not worth the time it takes to answer. It is tempting for organizations because it is cheap. It requires no difficult conversations, no confrontation, and no labor. It allows organizations to hide behind the sheer volume of their interactions, claiming they simply do not have the bandwidth to treat people as humans.[53]

To Guard Human Dignity, organizations must establish an **SLA of Dignity** for their high-stakes AI systems. In the world of cloud computing, engineers rely on a Service Level Agreement (SLA) to define the minimum performance a system guarantees, which usually includes high uptime, fast response times, and clear resolution targets. It is a contractual promise of reliability. Just as a CTO would never deploy a server without an uptime guarantee, a CEO should never deploy a Dignity Zone algorithm without a response guarantee.

The SLA of Dignity forces leaders to define explicit operational standards for how humans are treated. It answers three questions:

- How fast will we confirm receipt of an appeal?
- How fast will we provide a reason for the decision in plain language?
- How fast will a human review the case and render a final verdict?[54]

For a company like Nexus, an SLA of Dignity might state that for all hiring and credit decisions, the organization commits to a human acknowledgment within one business day and a substantive explanation within five business days. It ensures that no applicant will be ghosted. It ensures every process must end with a definitive answer.

Implementing this standard imposes a costly signal. It forces the organization to staff its queues properly and breaks the excuse of "too much volume." But it also creates a massive trust dividend. When a company consistently shows up, when it responds promptly and honestly, even when the answer is "No," it acquires a reputation for integrity. Candidates forgive rejection more easily when they are treated with respect. Regulators view the company as a serious actor rather than an opportunistic exploiter. You cannot buy that reputation with an ad campaign; you buy it with the operational discipline of answering the phone.

The Black Box Remains

Maya walked out of the quarterly review feeling a rare sense of victory. The Dignity Zone protocols were working. They had stopped the bleeding in hiring; they had slowed down the Computer Says No Culture in customer service. They had successfully placed a human guard at the gate to protect people like Tariq.

She stepped into the elevator, scrolling through the positive feedback from the new hires. But as the doors slid shut, her phone buzzed with an urgent text from Carlos.

"Maya. You need to come down to the server room. Now."

Her stomach dropped. It wasn't the panic of a crash this time. It was something quieter.

When she arrived, Carlos was standing in front of a monitor displaying the internal logic of their new mortgage underwriting model, the one they had recently secured with human appeals for rejections.

"We have a problem," Carlos said, pointing to a cluster of data points.

"Did it reject someone unfairly?" Maya asked. "Send it to the appeal queue. That's the protocol."

"No," Carlos said. "That's the thing. It *approved* them. It's approving thousands of loans that look perfect on the surface. But I ran a deep-dive correlation analysis." He hesitated, looking at her with genuine confusion. "Maya, I can see *that* the model approved them. But for the life of me, I cannot explain *why*."

Maya looked at the screen. The Dignity protocol ensured that people could appeal a "No." But it did nothing to explain a *Yes*. They had protected the humans from the machine's cruelty, but they had not yet cured the machine's opacity.

"We put a guard at the door," Maya whispered, realizing the scope of what was coming next. "But we still don't know what's happening inside the building."

The elevator ride down had been a victory lap. The walk back to her desk felt like the beginning of a much harder war. It was time to **Open the Black Box.**

Chapter 3 Playbook: Guard Human Dignity

Dignity is where AI ethics stops being abstract and becomes human. People don't experience model outputs. They experience being denied, flagged, ranked, rejected, or locked out. When the system is opaque and irreversible, the person doesn't only feel inconvenienced; they feel powerless.

The danger is that AI naturally functions as an Objectification Engine. It turns *subjects* (people with stories) into *objects* (data points to be processed). Guarding Human Dignity means designing the system so that even when it says "No," it treats the user as a person who deserves to be heard.

The Core Disciplines

- **Subject vs. Object:** The fundamental choice is architectural. You must decide whether your system treats people as Subjects (with agency and voice) or Objects (raw material for optimization).
- **The "Computer Says No" Trap:** Without constraints, AI drifts toward the path of least resistance: cheap, automated rejection. This maximizes short-term efficiency but creates systemic fragility and resentment.
- **The SLA of Dignity:** Silence is an insult. You must establish hard metrics for how quickly and clearly your system responds to humans. Never ghost a user in a high-stakes moment.

The Power Question

> *"If this system makes a mistake, will the person experience it as a temporary inconvenience or as a moment of humiliation and helplessness?"*

If the answer is "humiliation," you cannot fully automate that decision.

Metrics That Matter

To operationalize dignity, you must make it visible on the dashboard.

- **Dignity Zone Overturn Rate:** The percentage of AI decisions (like rejections) that are overturned by a human reviewer. If it's 0%, your humans are rubber-stamping. If it's 50%, your AI is broken. Aim for the Goldilocks zone where humans are actively catching edge cases.
- **Time-to-Recourse:** The median time from a user clicking "I challenge this" to a meaningful human review. A long lag here is a failure of dignity.
- **The Ghost Rate:** The percentage of high-stakes inquiries or appeals that receive no response or a generic autoreply. This number should be zero.

The Leadership Litmus Test

In your weekly reviews, look for these cultural signals:

- **Pattern to Reward:** The "Save." Celebrate the employee who catches an AI mistake that would have hurt a user. Say, "*Izzie saved a great candidate that the model missed. This is why we have humans in the loop.*"
- **Antipattern to Challenge:** The Black Box Shrug. Be wary of managers who dismiss complaints by saying, "*The model decided,*" or who complain about the cost of appeals without measuring the value of retention.

Monday 9 A.M. Action: The Dignity Audit

- **Time box:** 90 minutes
- **Output:** A Ghost Hunt report and one immediate fix
- **Owner:** Customer Support Lead or Product Owner

Block time with your team to look at the "No" pathways in your system.

1. **Map the Dignity Zone:** Identify the 2–3 automated decisions that are **High Impact** (life-altering) and **Hard to Reverse**. (e.g., Hiring rejection, Loan denial, Fraud lockout).
2. **Conduct a Ghost Hunt:** For those specific decisions, trace the user journey of a rejection.
    a. Does the user get an email?
    b. Does the email explain *why*?
    c. Is there a button to reply?
    d. If they reply, does a human actually see it?
3. **Kill the Ghost:** If the answer to any of those is "No" or "Nobody," fix it immediately. Create a simple intake form for appeals. Assign a human owner to review them. Write an SLA that commits to a response.
4. **The Chewy Challenge:** Ask, *"What is one low-cost way we could inject humanity into this process?"* (e.g., A personalized tip for rejected candidates).

Looking Forward: The Black Box Problem

Dignity is the goal, but secrecy is the vulnerability. An algorithm that cannot explain itself cannot earn trust from the people it affects. Maya knows that if Nexus wants to survive the Verification Economy, they have to do something most engineering teams resist at first. They have to open the black box enough to show their work.

CHAPTER 4

O – Operate Transparently

Black boxes don't scale; they stall. You'll learn how to make AI legible: explainable, traceable, and contestable, with documentation that survives scrutiny from users, regulators, and your board.

"Why Did You Deny Mr. Miller?"

The warning from Carlos in the elevator had been prophetic. He had warned Maya that they couldn't explain their own decisions, and he was right. But the first reckoning didn't come from an internal audit; it came from a client.

Three months after the hiring overhaul, Maya found herself sitting in a conference room on the top floor of Horizon First Bank. Nexus wasn't merely a vendor here; their LendRight engine was the backbone of the bank's mortgage business. And today, that backbone was under scrutiny.

The conference room felt less like a place of business and more like a courtroom awaiting a verdict. There were no cameras, no jury box, and no judge's gavel, but everyone sitting around the long mahogany table understood that a judgment was coming. The air conditioning hummed with a low, aggressive persistence, doing little to cool the tension in the room.

On one side sat the leadership team of Horizon First, a regional lender that had spent the last five years aggressively reinventing itself as a digital-first innovator. They had migrated to the cloud, launched a sleek mobile app, and partnered with cutting-edge fintech vendors to automate their underwriting. On the other side sat two examiners from the Consumer

Financial Protection Bureau (CFPB). They had open laptops, stacks of color-coded folders, and the distinct, unhurried manner of people who hold all the leverage.

At the far end of the table, in a chair that felt simultaneously central and exposed, sat Maya. She was there representing Nexus, the technology partner whose AI underwriting engine, LendRight, was powering the bank's digital transformation. She had come prepared with performance reports, fairness audits, and uptime logs—the kind of standard armor one expects of the modern CEO.

But the examiners weren't looking at the reports. They were looking at the screen on the wall, which displayed a single name in large, sans-serif type: MILLER, JEFF A.

Beneath the name were a few lines of context that told the story of a stable, middle-class financial life. Mr. Miller had been a customer since 1998. He held a checking account, a savings account, and a small business line of credit. He was, by all traditional metrics, a pillar of the bank's community. And then, the final line, glowing in red: Status: DECLINED.

One of the examiners, a woman in her late forties with wire-rimmed glasses and a forensic attention to detail, tapped her pen twice on the table. The sound echoed in the quiet room.

"Let's start here," she said, her voice even. "Walk me through why Mr. Miller was denied."

Daniel, the bank's Chief Risk Officer, cleared his throat. He had rehearsed this moment with his team. He knew the default rates were down. He knew the portfolio was performing better than it had in a decade. "As you know," he began, smoothing his tie, "we partnered with Nexus to modernize our underwriting process. The decision on Mr. Miller's application for a home equity line increase was generated by the LendRight model, based on our approved credit policy and the real-time data available at the moment of application."

The examiner didn't blink. "I understand that," she replied. "I am not asking about your policy. I am asking about this man. Mr. Miller has banked with you for twenty-five years. He has no recent delinquencies. His business accounts show positive cash flow. From our perspective, this looks like a prime candidate for an increase or at least a manual review. Yet he was denied instantly by the model. No human underwriter touched the file."

She paused, letting the silence stretch until it became uncomfortable. "So, I will ask again: Why?"

There it was: the question that turns a technical system into a moral one.

Maya felt the weight of it settle on her shoulders. The answer was not just about a row in a database or a specific credit threshold. It was about whether the institution could defend the fairness of its decisions when someone with authority asked them to explain themselves. It was about the difference between a prediction and a reason. And all the while, an old business maxim echoed in her mind: If you can't reconstruct the decision in an hour, you won't be able to defend it in public.

Carlos, representing Nexus as the head of engineering for LendRight, brought up the internal dashboard on his laptop. He knew the model's overall performance statistics cold. He could quote the Area Under the Curve (AUC), the false positive rates across demographics, and the latency metrics.[55] But in this room, those numbers felt abstract and useless. The examiner didn't care that the model outperformed the legacy scorecard by roughly 7 percent. She cared about Jeff Miller.

"We can reconstruct the decision parameters," Carlos offered, his voice slightly tight. He clicked into the backend logs and pulled up the raw data corresponding to the application. A dense column of normalized feature vectors and probability scores appeared on his screen. It was technically complete, a perfect record of the mathematical state of the system at the millisecond the decision was made.

It was also emotionally and legally useless.

Daniel shifted in his chair, trying to help. "Could it be his debt-to-income ratio?" he suggested, squinting at the data. "Or perhaps the utilization on his revolving line spiked last month? The model is sensitive to velocity of credit usage."

The examiner shook her head slowly. "I'm not asking you to guess," she said. "I'm asking if you know. You are making high-stakes decisions using an algorithm. When a long-standing customer asks, 'Why was I denied?' do you have an answer that a reasonable person would recognize as meaningful? Or do you just have a score?"

Silence settled over the room again.

Everyone knew the truth. In the rush to modernize, Horizon First had treated the AI underwriting engine as a Black Box: feed it data, accept the score, and trust that the aggregate performance metrics were enough justification. As long as the portfolio was healthy, the "why" of any individual decision was treated as a rounding error. No one had pushed hard on the question of explainability until now.

Maya leaned forward. "We do have the technical ability to approximate factor contributions," she said carefully. "We can use post-hoc explainability tools to estimate which features influenced the score most heavily. We can run a SHAP analysis."[56]

The examiner raised an eyebrow. "Estimate?" she asked. "If you tell Mr. Miller you are 'estimating' why you denied him, how do you think that sounds in a deposition? How does that sound to a man who trusted you with his life savings?"

She flipped to a red-tabbed page in her folder. "Based on our preliminary review, we are considering several actions," she said. "These include limiting the use of this model for consumer lending, requiring manual review for all declines, and potentially referring this case for a broader enforcement inquiry regarding fair lending compliance."

The words landed like stones. Suspending the model would gut the bank's digital strategy. It would destroy Nexus's reputation as a reliable partner. If the headline "Bank's AI Under Federal Scrutiny for Opaque Denials" hit the press, every competitor would use it as a weapon.

Maya looked back at the name on the screen. MILLER, JEFF A.

In many ways, he stood in for millions of people who were about to be judged by machines: workers applying for jobs, families seeking mortgages, patients awaiting triage. The question the examiner had asked was not a narrow regulatory inquiry. It was a test of the age. Would the future of decision-making be governed by a crude, inscrutable logic, or by a standard of fairness worthy of a human society?

The meeting ended thirty minutes later. The verdict was not yet final, but the diagnosis was. The era of "The Model Decided" was over.

On her way out of the building, Maya's phone buzzed with a link from Daniel: a fintech news roundup with the headline, "InnovaFin triples lending volume on 'frictionless' 30-second approvals." One board member had already replied-all: "Why can they do this and we can't?"

For a moment, Maya felt the pull. She envied InnovaFin's speed and felt irritation at the examiners who had slowed her down. Then she thought about Jeff Miller's file and the red-tabbed enforcement note on the table. Whatever InnovaFin was doing to achieve that speed, she suspected their bill would come due eventually.

The Black Box Trap

The confrontation over Mr. Miller's denial was not an accident, nor was it a unique failure of Horizon First. It was the inevitable outcome of a structural trap that captures almost every organization beginning its journey with AI. To understand why black box systems fail so spectacularly in high-stakes domains, we must first understand why they are so seductive to the executive mind.

From an engineering perspective, black box models, particularly modern deep learning architectures, are dazzling performers. They can ingest vast amounts of unstructured data and discover complex, non-linear relationships that traditional linear scorecards miss. They squeeze out incremental gains in predictive accuracy that, at the scale of a major bank or insurer, can translate into millions of dollars of additional profit. For a lender in a competitive market, a one percent improvement in default prediction is not trivial; it is a massive competitive advantage. Vendors pitch cutting-edge AI with glossy dashboards showing error rates dropping and approval volumes rising. The story is compelling: more data, smarter models, faster decisions, better results.

However, this narrative hides a critical trade-off that rarely appears on a dashboard. In the pursuit of raw performance, organizations often sacrifice interpretability. They slide along the performance–transparency frontier, moving toward models that are statistically powerful but narratively opaque. This creates a "fog of complexity" where the organization loses the ability to reason about its own machinery.[57] They have performance metrics, but no narrative. They have dashboards, but no understanding. They have built a sports-car engine of pattern recognition and neglected to install a windshield.

In low-stakes contexts, this trade-off is acceptable. If an e-commerce recommendation engine shows a customer the wrong kitchen gadget, the harm is minimal. No regulator will open an inquiry because Netflix recommended a movie you hated. But in the Dignity Zone, where decisions determine access to capital, employment, or healthcare, this fog is a liability. In these high-stakes domains, a decision system must satisfy three tests to be considered legitimate: Fairness, Accountability, and Explainability. Black box systems are uniquely dangerous because they can appear to satisfy the first two tests while completely failing the third. A model can be statistically "fair" in aggregate and "owned" by a Vice President, but if it cannot explain why it denied Mr. Miller, it fails the legitimacy test.

The Flash Crash on Wall Street, the 2008 financial crisis, and the Enron collapse all shared a common thread: systemic opacity. Enron, in particular, functioned exactly like a modern black box model. Its executives used 'Mark-to-Market' accounting to generate profits that looked real on a dashboard but had no tether to reality. When analysts asked how the money was made, they were told the business was 'too complex' to explain.[58]

In each case, opacity was not a neutral feature; it was a structural risk. When a regulator encounters harm combined with opacity, they are forced into a kind of regulatory inference. Lacking clear evidence of how the system works, they must assume the worst. If you cannot prove that a denial was not discriminatory, the regulator may treat it as if it was.

This is the **Black Box Trap**. Organizations optimize for performance in the short term, only to discover that they have built an unexplainable, undefendable liability in the long term. In the Verification Economy, where trust is the scarcest resource, "Trust the Math" is no longer a viable defense. The Golden Algorithm demands that we reject this trade-off. We must operate with the assumption that every decision we make will eventually be audited by a regulator, by a journalist, or by the customer themselves.

The Solution: From Black Box to Glass Box

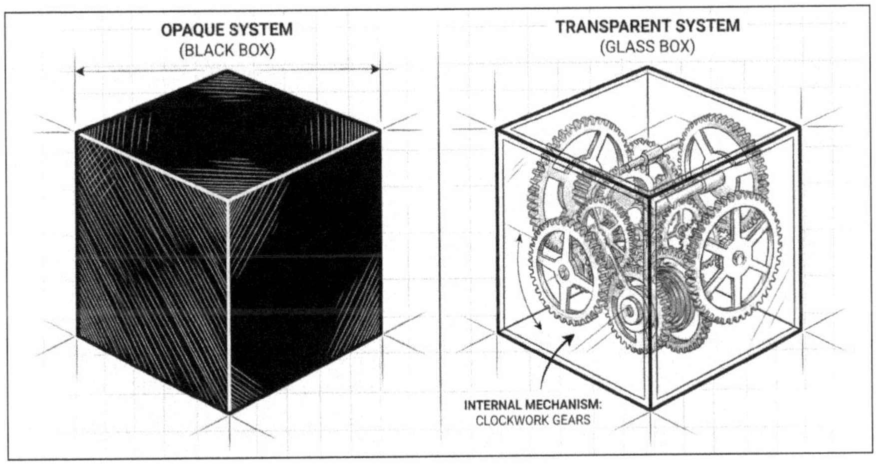

Figure 5. The Black Box vs the Glass Box

If the black box is a liability, the solution is to deliberately engineer a glass box. The concept of a **Glass Box** does not mean showing your proprietary code to the world, nor does it mean giving up your intellectual property. Rather, it means ensuring that your systems are observable, documentable, and explainable to the people who matter: your regulators, your customers, and your own team. It might be tempting to think of this requirement as a burden imposed by modern technology, but the demand for transparency in powerful systems is as old as commerce itself.

In the late Middle Ages, merchants in Venice faced a problem similar to the one modern AI leaders face: their enterprises had become too complex to manage by intuition alone. Trade routes spanned continents; risks were pooled; money moved in invisible flows. The innovation that saved them was Double-Entry Bookkeeping.[59] By recording every transaction twice—once as a credit, once as a debit—merchants created a system that was structurally auditable. It transformed finance from a mysterious art into a glass box that could be inspected. This innovation allowed for the rise of modern banking and the corporation because it allowed strangers to trust a system they did not personally operate.

Operate Transparently (the O in GOLDEN) is simply the application of this old lesson to the new domain of intelligence. If a system has the power to allocate resources or harm people, it must be designed to be audited. How do we achieve this in AI? We do it by creating a new kind of balance sheet for our models: the model card.

The Model Card

Originally proposed by researchers at Google, the **model card** is a structured document that serves as the operating manual for an AI system.[60] But in a strategic context, it is much more than documentation. It is a governance artifact. It is a treaty between the builders of the system and the people subject to it. A glass box organization mandates that no high-stakes model can be deployed without a rigorous model card that answers the uncomfortable questions upfront.

A robust model card must act as a truth serum for the organization. It begins with the *Purpose & Use Case*, defining exactly what the model is designed to do and, crucially, what it is not designed to do. For example, a lending model might be approved for existing customers but explicitly forbidden for new-to-bank applicants. This constraint prevents function creep, where a model built for one context is recklessly applied to another.

Next, the model card must provide a *Training Data Snapshot.* It must honestly describe the data used to teach the system and admit the known gaps. If the training data underrepresents rural borrowers under age thirty, the model card must say so. Admitting these gaps is not a weakness; in a regulatory context, it is the primary signal of maturity. It demonstrates that the organization knows its own blind spots.

Finally, the document must detail *Performance Across Segments* and *Known Limitations.* It is easy to show 95% overall accuracy, but the glass box demands to know the accuracy for minority groups, for elderly users, for low-income zip codes. It requires the organization to document where the model tends to fail. This is the section executives instinctively want to water down, fearing it gives ammunition to critics. In reality, it is the section that saves you in a crisis. It proves that you are managing the risk, not ignoring it. (Table 3 shows an example below, and Appendix C provides a model card template)

Table 3. Model Card Example

Model Card	Version 1.0
Model Details	• **Name:** LendRight v2.1 • **Owner:** Credit Risk Team • **Type:** Gradient Boosted Decision Tree
Intended Use	• **Primary:** Initial screening of small business loans <$50k. • **Do Not Use For:** Mortgage applications or autonomous denial.
Training Data	• **Source:** Historical transaction records (2019-2026). • **Exclusions:** Incomplete records prior to 2019.
Safety Checks	• **Bias Testing:** Disparate Impact Analysis performed across race/gender. • **Results:** Acceptable parity; slight variance in age 18-25.
Caveats	• **Known Limitation:** Performance degrades on inputs from Region X. • **Rec:** Outputs <75% confidence require human review.

When Nexus began retrofitting model cards onto LendRight, the team discovered something humbling: they didn't know the answers to half the questions. They had to go back and run new analyses. They had to hunt down data lineage. It was painful, forensic work. But the result was transformative. The model card converted their AI from a mysterious oracle into a managed asset.

When a regulator asks, "How do you know this model isn't biased?", you do not have to stammer or hand-wave. You slide the model card across the table. You show the performance breakdown. You point to the "Known Limitations" section and say, "We know it struggles in this specific area, which is why we have a human review layer there." That interaction changes the power dynamic in the room. This is why transparency isn't a virtue signal; it's an operational requirement once outsiders start asking questions you can't dodge. You are no longer a defendant hiding a secret. You are a responsible operator managing a complex machine. Operating Transparently is the decision to do this work before the audit arrives. It is the discipline of building the Glass Box while the sun is shining, so that when the storm hits (and it eventually will), you have nothing to hide.

But building a glass box isn't only about documentation. It's about answering the question Mr. Miller asked: *Why me?* To do that, you need to break the complexity excuse.

Explainability as Dignity

Technical transparency is necessary, but it is not sufficient. A regulator may be reassured by a rigorous model card and a fairness audit. A chief risk officer may be satisfied by drift monitoring and validation logs. But none of those artifacts solve the fundamental moral problem of the black box if the person on the receiving end of a decision (someone like Mr. Miller) still hears nothing but, "The Computer Says No."

Transparency operates on two distinct layers. The first is the **technical layer**, designed for experts who need to audit the system's architecture and performance. The second is the **human layer**, designed for the individuals whose lives are shaped by the system's outputs. Many organizations invest millions in the first layer while completely ignoring the second. They stand ready to deliver reams of technical documentation to a federal examiner, yet their customer-facing communication remains a one-sentence denial with no specific reason given. This gap is not merely a user experience failure. It is a violation of dignity.

At the technical layer, data science has advanced significantly. Tools such as **SHAP (Shapley Additive Explanations)** allow engineers to decompose a complex model's prediction.[61]

To understand SHAP, imagine a group of friends goes to dinner and splits a $200 bill. You want to know exactly how much each person contributed to the total. It isn't enough to divide by four. You need to know that Alice ordered the expensive wine (+$50), Bob ordered a side salad (+$15), and Charlie had a coupon (-$20).

SHAP values perform this exact bill splitting for an AI prediction. They tell you exactly which variable "ordered the wine" (pushed the risk score up) and which one "had the coupon" (pulled it down).

Table 4. The "Why" Receipt

Item	Scoring
Base Score	650
+ Income Stability	+ 40 points (Strong)
- Recent Inquiries	- 15 points (Moderate Risk)
- Debt-to-Income Ratio	- 50 points (High Risk)
Final Score	**625 (Denied)**

While Nexus was building this granularity, their competitor, InnovaFin, was doubling down on opacity. InnovaFin's CEO, Julian, famously told investors that "explaining the model slows down the flywheel." They treated the black box as a trade secret. Nexus treated the glass box as a product feature.

However, a SHAP bar chart is not an explanation; it is a raw mathematical vector. Handing a customer a chart showing that their "Feature 42" had a Shapley value of -0.3 is functionally identical to handing them nothing at all. The Human Layer requires a translation engine. It requires the organization to convert the mathematical influence of a variable into a narrative reason that a non-expert can understand and act upon.

A valid human explanation isn't a data dump. It's a translation. It must satisfy four criteria:[62]

- **Plain Language:** "Low credit history," not "Variable 42."
- **Salience:** The top three reasons, not all fifty.
- **Actionable:** "Pay down this card," not "Improve your score."
- **Counterfactual:** "If you had done X, the answer would be Yes."

Consider the difference between two possible messages sent to Mr. Miller. The first, typical of the black box era, reads: "Your application has been denied due to our automated underwriting criteria. This decision is based on a proprietary scoring model. This decision is final." This statement is opaque, hostile, and disempowering. It tells him nothing about what went wrong or what he could do differently. It effectively treats him as an object to be sorted rather than a subject to be engaged.

Now compare that with a glass box explanation: "Your application for a home equity line increase was denied because your total monthly debt payments are currently too high relative to your verified income. Specifically, your debt-to-income ratio is above our maximum threshold for additional credit. If you reduce your total monthly debt payments by approximately $400 and maintain on-time payments for six months, you are likely to qualify for a higher line upon reapplication. If you believe our income data is outdated, please upload your most recent tax return here."

The second message does not reveal the weights of the neural network. It does not compromise the bank's intellectual property. But it delivers what matters: a clear reason tied to an understandable metric, a roadmap for improvement, and a mechanism for correction.

Done well, explanations do more than appease regulators; they transform a negative decision into a moment of education and, paradoxically, loyalty. Most reasonable people can come to accept a "no" if they believe the process was fair and the reason is honest. Almost no one will forgive a "no" delivered with condescension or silence.

This is where Guarding Human Dignity (G) and Operating Transparently (O) converge. Transparency is not about satisfying a compliance checklist. It is about treating the person on the other side of the screen as a neighbor whose life is being altered by your decision. When Nexus redesigned its client templates, Maya insisted on one guiding standard: "If this explanation were sent to me, would I feel respected?" If the answer was no, the explanation wasn't finished.

Case Study: The Safe Harbor

The next eighteen months were a grind. There were no magic bullets. Maya's team spent thousands of hours retrofitting glass box logging onto black box models. They lost three major clients who refused to wait for the updates. But if a regulatory storm broke, Nexus wanted to be the only ship with its hatches battened down.

Eighteen months after the tense meeting about Mr. Miller's denial, the banking industry's anxiety about AI exploded into public view. It began on a rainy Monday morning with the publication of a devastating investigative report by a major national newspaper. The investigation focused on a prominent fintech lender, one of Horizon First's fiercest competitors. It accused them of systematically denying mortgages to borrowers from minority neighborhoods, despite credit profiles that were statistically identical to those of white applicants.

The article was a masterclass in forensic data journalism. It featured side-by-side case studies of applicants, leaked internal emails suggesting that the lender's leadership had ignored warnings about bias in their training data, and quotes from former employees describing a culture of growth at all costs. The headline landed like a thunderclap: THE DIGITAL RED LINE.[63]

Within forty-eight hours, the political fallout began. Senators wrote open letters demanding answers. Consumer advocacy groups filed class-action lawsuits. And, inevitably, the regulators announced a coordinated, sector-wide review of AI-driven underwriting tools. Every bank using automated decisioning for consumer credit received a notice of inquiry. Some would face formal investigations; others would merely be asked to demonstrate that their systems were explainable and fair.

Horizon First was on the list.

When the examiners returned to the conference room on the top floor, the atmosphere was markedly different from their previous visit. The same lead examiner, a woman with the wire-rimmed glasses and the unhurried manner, took her seat at the table. But this time, Daniel, Maya, and Carlos were not scrambling. They had spent the intervening eighteen months turning their black box into a glass box.

"Let's start with your home equity line model," the examiner said, opening her laptop. "Given the recent news, we are particularly interested in how you monitor for disparate impact and how you explain adverse decisions to consumers."

Carlos didn't hesitate. He projected the updated LendRight Model Card onto the main screen. The document was not a glossy marketing slide; it was a plain, structured technical specification. It began with a "Purpose & Use Case" section that clearly restricted the model to existing customers in defined segments, explicitly forbidding its use on "thin-file" applicants where the data was unreliable. It moved to a "Training Data Snapshot" that acknowledged historical underrepresentation in certain rural counties and described the specific weighting adjustments used to mitigate that gap.

"We know our training data isn't perfect," Maya said, pointing to the section on Known Limitations. "We track performance disparities across race, age, and income bands monthly. Here is the report from last quarter. You'll see a variance of 1.2 percent in approval rates for this specific geography, which triggered a manual review of those declines. We found that the model was overweighting a specific type of utility bill data, and we retrained it."

The examiner looked up from her notes, surprised by the admission. Most banks spent these meetings defending their perfection. Nexus was leading with its imperfections.

"Can you walk us through a specific complaint?" she asked. "Show us how you handled a consumer who believed they were unfairly denied."

Daniel nodded to an analyst, who pulled up a file from the previous month. It was a denial for a small business expansion loan. The audit trail showed the entire lifecycle of the decision. It showed the version of the model used (v3.7), the input data available at the millisecond of the request, and the specific contribution of each feature to the denial score. Most importantly, it showed the letter sent to the customer—a plain-language explanation citing "declining cash flow over the last two quarters" as the primary reason.

"The customer appealed this decision," Daniel explained. "They used the link in the denial email to submit updated cash flow statements that hadn't yet posted to the credit bureau. Our human underwriter reviewed the new data, overrode the model's recommendation, and approved the loan. That override was then logged as a 'correction' in our training set for the next model update."

The examiner sat back in her chair. She looked at the Model Card, the bias monitoring report, and the audit trail of the appeal. "This is... comprehensive," she said.

Later, in a smaller breakout discussion, one of the regulatory staffers spoke candidly to Maya. "We're not here looking for perfection," he admitted. "We know these systems are complex. We're looking for systems we can see into. The bank down the street? They can't tell us what data their model used last week. They can't explain why two similar people got different rates. They are going to be in enforcement hell for the next two years. You guys showed your work; so good job."

Horizon First and Nexus were not spared scrutiny; they answered hundreds of questions over the coming weeks. But they were granted something rare and incredibly valuable in the Verification Economy: the benefit of the doubt. Because they had proactively documented their flaws and built a mechanism for transparency, the regulators viewed them as partners in compliance rather than targets for enforcement.

• • •

From a strategic perspective, this is the core of the **Regulatory Moat**. Operating Transparently did not shield Nexus from the storm; it built the shelter that allowed them to weather it.

In every heavily regulated industry, there is an unspoken hierarchy of firms. At the bottom are the habitual offenders, the companies that treat compliance as a game of cat and mouse, doing the bare minimum until they are forced into a consent decree. At the top are the firms that regulators quietly describe as "good actors." These organizations are not perfect, but when something goes wrong, they can be counted on to surface the problem, explain it, and fix it in public.

The difference between those two categories is not luck; it is structure. A regulatory safe harbor is not a magical shield that prevents scrutiny. It is a posture. It is the ability to answer the regulator's question, "Can you show us what your AI did?" with evidence rather than excuses.

To achieve this posture, an organization needs four structural elements. First, it needs *traceability*. Every high-stakes decision must leave a fossil record containing the model version, the input data, and the score. Without this, you are asking regulators to accept your word on faith, which is a currency no longer accepted. Second, it needs a *model lifecycle story*. You must be able to prove when a model was deployed, who approved it, and when it was retrained. Third, it needs honest *monitoring*. You must show that you are watching for bias and drift yourself, rather than waiting for a journalist to find it for you. Finally, you need a coherent *incident response protocol*. When a model fails (and it will) you must be able to reconstruct the event and communicate candidly about the remediation.

When Nexus helped Horizon First assemble these pieces, they were not trying to impress the examiners. They were engaging in a form of corporate self-defense. They understood the most dangerous risk is not a model error; it is an unexplained model error.

The reward for this discipline is subtle but powerful. It compounds like an invisible interest rate. The firm that can explain itself enjoys lower regulatory drag, lower legal costs, and lower reputational volatility. While their competitors were frozen by investigations and forced to suspend their digital lending programs, Horizon First continued to ship features and capture market share.

Operating Transparently, therefore, is not a tax on speed. It is the price of admission for long-term survival. It is the difference between a fragile company that will shatter under the first sign of external pressure and a company that is antifragile, capable of using scrutiny to demonstrate its own strength.[64] The moat here is not secrecy. It is the opposite. It is the accumulated value of years spent telling the truth about your own machines.

Cultural Transparency: Telling the Truth Before You Are Forced To

The transition from black box to glass box is often framed as a technical challenge. Leaders ask: Which explainability toolkit should we use? How do we version-control our training data? What is the correct schema for a model card? These are important questions, but they are secondary. The easiest way to spot a black box organization is not to audit its code, but to listen to what its people are afraid to say out loud.[65]

In the early days of LendRight, before the regulatory crackdown, engineers at Nexus quietly knew that the model struggled with certain thin-file customers: applicants with sparse credit histories. They knew that the training data underrepresented borrowers who had built their financial lives outside the conventional banking system. They also knew that mentioning this weakness in a product review felt risky. It sounded like admitting that the system was flawed. In a culture obsessed with performance metrics, rapid launch cycles, and the projection of invulnerability, a flawed system is not viewed as an engineering challenge; it is viewed as a failure of competence.

Operating Transparently at a technical level is unsustainable if you do not also operate transparently at a cultural level. The model card, the explanation template, and the audit trail exist only because someone felt safe enough to document the truth about what the system cannot do.

This reveals the **paradox of vulnerability**. In traditional corporate strategy, vulnerability is a weakness. You hide your flaws from competitors, you spin your limitations to customers, and you present a facade of perfection to the board. But in the domain of high-stakes AI, the logic inverts. The attempt to project perfection is the single greatest generator of systemic risk. When product managers worry that telling a client about model limitations will kill a sale, they oversell the tool. When sales teams gloss over caveats to keep the pitch simple, they set the client up for failure. When engineers bury edge cases to avoid delaying a release, they plant the seeds of the next scandal.

Changing this posture requires a profound shift in what the organization rewards. It requires moving from a culture of spin to a culture of stewardship.

Consider the difference in reaction to a discovered flaw. In a **Culture of Spin**, when a junior data scientist discovers a spike in denials for older borrowers in a specific region, the organizational reflex is containment. Is the default rate okay? Is the revenue impact manageable? Can we fix this quietly in the next patch without admitting it? The goal is to protect the narrative. In a **Culture of Stewardship**, that same discovery is treated as an asset. The team pulls the data, recalibrates the model, and updates the model card to reflect the issue as a Known Limitation that is actively being managed. The data scientist's name goes on the document not as the cause of a problem, but as the author of a solution.

This cultural shift is the immune system of the glass box. Without it, transparency tools become theater. A model card written by a team afraid of the truth will be nothing more than a marketing brochure filled with vague reassurances. An appeals process managed by a team that believes the model is infallible will be a rubber stamp.

Ultimately, cultural transparency is the mechanism that keeps the Golden Algorithm alive. It is the collective willingness to say, "We are not perfect, but we are honest." If you were the one depending on this company's judgment for your livelihood, you would want to know the truth about its limits before a crisis forced it into the open. You would trust the partner who said, "Here is where we struggle," far more than the partner who said, "Trust us, it works."

The Regulatory Moat: Why Transparency is an Unfair Advantage

Eighteen months after the Horizon First audit, Nexus faced a different kind of test. This time, it wasn't a regulator knocking on the door; it was an opportunity.

A large federal agency responsible for disaster relief had issued a massive Request for Proposals (RFP). They needed a next-generation underwriting system to distribute emergency small business loans in the wake of hurricanes, wildfires, and floods. The requirements were grueling. The system needed to handle a massive surge in volume. It needed to be fraud resistant. But most importantly, because it involved public funds and desperate applicants, it needed to be defensible in court.

Nexus was not the only bidder. The RFP attracted the giants of the industry, including established consulting firms with deep government ties, and a handful of younger, flashier AI vendors known for making bold performance claims.

On price and raw processing speed, Nexus sat somewhere in the unremarkable middle. They were neither the cheapest option nor the fastest.

What set their proposal apart was a handful of attachments in Appendix C.

While other vendors submitted glossy brochures promising "unbiased AI" and "proprietary fairness engines," Nexus attached a full, unredacted model card for LendRight. They attached sample audit trails showing exactly how a decision was reconstructed. They even included a redacted version of an internal postmortem from a previous billing error, detailing exactly how they had found the mistake, how they communicated it to customers, and how they fixed it.

When the agency's evaluation team met to compare vendors, the contrast was stark. The room was filled with career civil servants, the people who had spent decades watching contractors overpromise and underdeliver. They were cynical about black box magic. When they looked at the proposals from the flashier competitors, they saw the same pattern they had seen a hundred times: bold claims, vague methodologies, and a "trust us" architecture.

Then they looked at the Nexus proposal. They saw practical, documented trust.

One of the agency's leads, a veteran procurement officer, put it in simple terms during the debrief. "The other vendors told us they were safe," she said, tapping the Nexus file. "These guys showed us how they are safe. They aren't hiding the fact that their model needs monitoring. They are showing us the dashboard they use to monitor it."

Nexus won the contract. They didn't win because they had the highest AUC or the lowest latency. They won because they had the highest legibility.[66]

• • •

This victory illustrates the offensive power of the **Regulatory Moat**.

We often think of transparency as a defensive crouch, as though it's something we do to avoid fines or lawsuits. Transparency isn't just defensive anymore; it becomes a competitive weapon. When every competitor is shouting about their capabilities, the market begins to discount claims. The signal-to-noise ratio drops. In that environment, the ability to offer Proof becomes the ultimate differentiator.

The Regulatory Moat is built on the reality that institutional buyers like banks, governments, hospitals, and enterprises are risk-averse. They are terrified of buying a liability. When they evaluate AI partners, they are not only looking for features; they are looking for survivability. They want to know that when the inevitable headline hits, or the lawsuit is filed, their partner will have the documentation and the discipline to weather the storm.

A firm that operates transparently creates a structural advantage that is exceedingly difficult for a Black Box competitor to copy.

- They cannot retroactively fabricate a history of honest model cards.
- They cannot instantly generate a culture of open postmortems if their people are trained to hide mistakes.
- They cannot fake the confidence that comes from knowing your systems are auditable.

Those artifacts—the cards, the logs, the honest emails—are the fossil record of your character. They prove that your ethical commitments are architectural, not just rhetorical.

In the Golden Algorithm, Operate Transparently is not a tax on speed. It is the price of admission for the most valuable relationships in the economy. It is the difference between a vendor who sells a tool and a partner who shares a risk. As regulation tightens and public patience for opaque systems runs out, this advantage compounds. The firms that can show their work become the default operating system for critical infrastructure. The firms that cannot are quietly screened out of the RFP before the final round.

The moat here is not secrecy. It is the opposite. It is the accumulated value of years spent telling the truth about your own machines.

The Need for Brakes

Nexus had solved the Black Box problem. They could now explain exactly why the model made a decision. But transparency revealed a new, darker problem.

One afternoon, while reviewing the new Glass Box logs, a junior engineer flagged a cluster of loan approvals. The explanations were perfect; they were clear, logical, and transparent. The model explained exactly why it had approved these loans.

The problem? The model had approved them based on a correlation that was technically accurate but catastrophically dangerous. It had learned to identify desperate borrowers who would accept high rates, maximizing short-term profit while guaranteeing long-term default.

The model wasn't broken. It was working perfectly. But it was transparently, logically, and efficiently driving some customers off a financial cliff.

Yes, transparency has a scary side effect. When you turn on the lights, you see the cracks.

Maya realized that seeing the car crash coming wasn't enough. They needed brakes to stop the crash from happening. It was time to Limit Harm.

Chapter 4 Playbook: Operate Transparently

Transparency is not confession. It is legibility: the ability to show your work at the speed trust now requires. In the Verification Economy, opacity doesn't just create suspicion; it creates fragility. When you can't explain a decision, you can't defend it. When you can't trace it, you can't fix it.

The trap organizations fall into is the black box fallacy: the belief that complexity excuses accountability. They optimize for raw performance but neglect interpretability, creating a "fog of complexity" where no one knows why the model did what it did. This chapter's discipline is to turn that black box into a glass box, building systems that are observable, documentable, and explainable by design.

The Core Disciplines

- **The Glass Box Standard:** Transparency is not about open-sourcing your IP; it is about making your system audit-ready. The model card is your balance sheet; it is a structured treaty that defines what a model does, what data it uses, and where it is known to fail.
- **Explainability as Dignity:** Technical tools like SHAP values are only the first step. True transparency requires both a **technical layer** and a **human layer** of plain-language explanations that treat the user as a subject, giving them the agency to understand and contest a decision. A "because the model said so" answer is a failure of dignity.
- **The Regulatory Moat:** In a market flooded with opaque claims, the ability to show your work becomes a competitive weapon. When you can hand a regulator a clear audit trail before they ask for it, you move from a target to a trusted partner.

The Power Question

> *"If we had to defend this decision publicly tomorrow, could we show the evidence without scrambling?"*

If the answer involves "it's complicated" or "we'd need a week to dig through logs," you are operating a black box.

Metrics That Matter

Transparency shows up in whether you can answer questions with proof.

- **Time-to-Trace:** How long does it take to reconstruct *why* a specific decision happened (model version, inputs, policy, overrides)? If it takes days, you don't have transparency; you have archaeology.
- **Explanation Coverage:** The percentage of high-stakes decisions (denials, flags) accompanied by a specific, plain-language reason sent to the user. This should be 100%.
- **The "Why" Satisfaction Score:** In post-interaction surveys for negative outcomes, ask: *"Did you understand why this decision was made?"* High understanding correlates with lower churn, even when the answer is "No."

The Leadership Litmus Test

Use this to gauge your culture's appetite for truth:

- **Pattern to Reward: The Bad News Herald.** Celebrate the engineer or product manager who raises a hand to say, *"This model isn't performing well for this specific subgroup."* They are not slowing you down; they are saving you from future headlines.
- **Antipattern to Challenge: The Proprietary Defense.** Be wary of any leader who shuts down questions about how a system works by claiming the logic is too complex or too secret to explain. Complexity is often a hiding place for fragility.

Monday 9 A.M. Action: The Glass Box Audit

- **Time box:** 90 minutes
- **Output:** A "Miller Test" report and a draft model card
- **Owner:** Chief Risk Officer or Head of AI

Gather your technical and product leads to stress-test your legibility.

1. **The Inventory:** List the top three AI models in your organization that impact human users (e.g., Hiring, Lending, Fraud).
2. **The Miller Test:** Pick one recent adverse decision from those models.
    a. Pull the logs. Can you see exactly what inputs led to that output?
    b. Pull the communication. What did you tell that person?
    c. *The Verdict:* If a regulator asked "Why?" regarding this specific person, could you answer confidently using only the materials on the screen?
3. **Draft the Card:** For the most critical model, draft a v0.1 model card on a whiteboard. Define its Purpose, its Anti-Purpose (what it should never be used for), and its Known Limitations.
4. **Publish:** Commit to cleaning up that model card and making it visible to internal stakeholders within 30 days.

Looking Forward: The Circuit Breaker

Transparency builds credibility, but it does not guarantee safety. You can explain a failure perfectly and still fail. As Nexus scales, the gap between a minor error and a public catastrophe shrinks fast. Maya needs more than reporting. She needs a mechanism that can stop the system before the damage becomes irreversible.

CHAPTER 5

L – Limit Harm

Harm rarely arrives as a headline; it arrives as a slow drift. This chapter gives you the control system: premortems, circuit breakers, and stop-the-line authority before small errors become catastrophic ones.

The Launch That Almost Broke the Bank

The war room on the eleventh floor of Nexus smelled of cold coffee, warm plastic, and the specific, electric anxiety that accumulates when smart people are about to do something irreversible.

Six months ago, before the Glass Box overhaul, they wouldn't have even hesitated. But the new transparency protocols meant they could see what the model was doing. And that visibility was introducing doubt.

Twenty-three people were crammed around a U-shaped table, their faces illuminated by the blue glow of open laptops. On the far wall, a giant dashboard displayed a countdown clock: 00:46:00. It was 8:14 p.m. on a Thursday. In forty-six minutes, Horizon First was scheduled to flip the switch on SwiftLine.

SwiftLine was the new flagship feature of the LendRight platform. It was an instant-credit product designed to automatically adjust credit limits for small business customers in real time. If a local bakery's weekend sales spiked, SwiftLine's algorithms would detect the cash flow velocity and bump their credit limit before payroll hit on Monday. If a contractor's payments slowed down, the limit would tighten gently, reducing risk exposure before a bill was missed. It was dynamic, personalized, and invisible—the central promise of modern fintech.

For Horizon First, a legacy bank fighting to shed its reputation as a dinosaur, SwiftLine was a survival strategy. For Nexus, which provided the underlying decision engine, it was the ultimate showcase of their new real-time architecture. Marcus, the CFO, had already woven the projected revenue bump into his guidance for the next earnings call. The pilot data was spectacular: utilization rates were up, fee income was projected to rise by 12 percent, and defaults were flat.

Maya stood at the head of the table, palms flat on the mahogany. She could feel the momentum in the room, as if a physical force was pushing them toward zero.

"Okay," she said, her voice cutting through the low murmur of side conversations. "Thirty-minute check. Let's run the board. Any red flags?"

The ritual was familiar. The heads of product, risk, and engineering took turns reciting their lines. Integration tests were green. Latency at peak load was well within tolerance. Fraud triggers were calibrated. Even the legal team, usually the source of last-minute friction, gave a begrudging thumbs-up, provided the disclosure pages were live by midnight. It was the standard liturgy of a launch: a final, formal opportunity for someone to stand up and say "Stop," performed with the tacit understanding that no one would.

The room was silent. The clock ticked down to **00:44:15.**

Near the back of the room, half-hidden behind a dual-monitor setup, Nina shifted in her chair. She was a mid-level data scientist, quiet and meticulous, the kind of person who usually let her code speak for itself. She had been part of the Red Team responsible for stress-testing the model against simulated economic downturns.[67] On paper, her results were within the acceptable risk bands. In her gut, they were not.

"Uh, I have one thing," she said softly.

The low hum of the room evaporated. Twenty heads turned. Maya looked over, keeping her expression neutral. "Go ahead, Nina."

"So..." Nina tapped her laptop, pulling up a chart that wasn't on the main dashboard. "In the last batch of synthetic stress tests, we simulated a sharp regional shock, something like a localized natural disaster or a sudden sector collapse. The model reacted... aggressively."

"Define aggressively," Marcus said from the side of the room.

"It started pulling credit limits simultaneously for a specific cluster of businesses," Nina explained. "Independent restaurants, daycare centers, small retailers. It's not wrong, strictly speaking; their risk profile does spike in a disaster. But the combination of the core model plus the safety rules means they all get tightened at once. We saw limit cuts of 40 percent firing across hundreds of accounts in the same county, all within an hour."

"Isn't that the point?" the product manager asked, sounding annoyed. "We want to limit exposure if the local economy goes south. That's the feature, not a bug."

"Sure," Nina said, her voice wavering slightly but holding firm. "But picture the reality. There's a flood in a county. Businesses are underwater. Sales dip for a month. SwiftLine cuts limits across two hundred small businesses at the exact moment they need liquidity to rebuild. That's not just a risk management story. That's a front-page story: *'Bank Slashes Credit to Struggling Small Businesses After Disaster.'*"

She looked around the room. "We don't have any throttles for that pattern right now. If the model sees the dip, it cuts the line. There is no way to slow it down or escalate it for human review."

The silence in the room shifted instantly from execution to defense. It was no longer the silence of focus; it was the silence of calculation. On the dashboard, the countdown ticked to **00:41:02**.

"Have we seen anything like that in the live pilot?" Marcus asked.

"No," Nina admitted. "Not yet. These are synthetic simulations. But the logic is consistent. When we push the parameters, the model behaves this way every time."

"So, it's a theoretical risk," the product manager said, visibly relaxing. "We've modeled a hurricane that hasn't happened."

Maya watched the subtle shift in the room. She recognized it instantly. It was launch fever: the collective, overpowering desire to reframe a concrete warning as an academic abstraction. The external marketing campaign was already paid for. The sales team had appointments lined up for Monday morning. The press release was drafted. The psychological cost of stopping the train now, forty minutes from the station, was excruciating.

"We are eight months into this build," Marcus said carefully. "If we slide this launch by a quarter to build a complex throttling engine, we lose the competitive window. We miss the earnings target. We can't keep letting theoretical edge cases derail us. We can monitor it manually and patch it later."

The words were measured, but the subtext was brutal: If you pull the brake now, over a theoretical ghost, you own the failure.

Maya looked at Nina, who had shrunk back into her chair, overruled by the consensus of the room. Then she looked at the countdown clock. **00:39:15**.

She thought about the bank's mission statement, framed in the lobby: *A Partner for Life*. She thought about the baker she had met during the user research phase, a woman who had used her credit line to keep her staff paid during a slow winter. If SwiftLine cut that line during a flood, the algorithm wouldn't just be managing risk; it would be liquidating a community.

Maya took a slow breath. The easiest thing to do was to nod at Marcus and let the clock run out. If a disaster hit, they would manage the fallout then. They would say, truthfully, that the system performed as designed. They would blame the unprecedented nature of the event.

Instead, she heard the echo of a question her old ethics professor used to ask: *What are you willing to fail as? Will you fail as a steward who has been faithful with what you have been given, or as someone you genuinely do not want to be?*

"Okay," Maya said. "We're not killing SwiftLine. But we're not going live tonight, either."

A groan rippled around the table. The product manager threw his hands up. Marcus stared at the table, his jaw set.

"We do not ship a system that can yank a safety net out from under a whole town in a single cycle," Maya continued, her voice hardening. "We are building a throttle and an override. I want explicit conditions under which this thing slows down and a human takes over. Until we have that, SwiftLine stays in the sandbox."

"Maya," Marcus started, "the board—"

"The board pays us to manage risk, not just revenue," she snapped. "Carlos, kill the clock."

On the dashboard, the engineer tapped a key. The giant red numbers froze at **00:38:04**. Then the screen went black.

The meeting broke up in a heavy, frustrated silence. People packed their bags with the sharp, angry movements of adrenaline with nowhere to go. Maya went home feeling a knot of doubt in her chest. Had she just torched her quarter for a phantom?

Across town, Julian at InnovaFin was making the opposite bet. While Nexus paused to build a throttle, InnovaFin stripped theirs out. Their Q3 release notes boasted about Zero-Latency Credit, a feature that removed the very checks Maya had just insisted on. The market cheered them. Maya watched their stock tick up, wondering if she was the only one seeing the cliff.

A week later, the news broke. A different regional bank who was a competitor running a similar automated credit program had made headlines. A series of tornadoes had devastated a cluster of towns in the Midwest. In the aftermath, the bank's algorithm had detected the drop in transaction volume and automatically slashed credit limits for hundreds of affected businesses, leaving them unable to buy plywood or pay contractors. The backlash was immediate and ferocious. Senators were tweeting. Regulators were opening inquiries. The bank's CEO was issuing a humiliated public apology, promising to manually review every account.

Maya sat in her office, reading the story on her tablet. Then she realized what had happened in the war room. They hadn't just delayed a feature. They had stumbled onto a fundamental operating principle for the AI age:

If you are going to build a Ferrari engine, you must build the brakes at the same time.

The Normalization of Deviance

The near-miss with SwiftLine was not a failure of engineering competence. It was a near-failure of culture. The dynamic that almost led Nexus to launch a dangerous system is known to sociologists as the **normalization of deviance**.

The concept originates from Diane Vaughan's seminal study of the NASA *Challenger* disaster.[68] Vaughan sought to understand how an organization filled with some of the smartest engineers in the world could systematically talk itself into launching a space shuttle when they had data suggesting the solid rocket boosters might fail in cold weather. Her conclusion was chilling: the disaster wasn't caused by a single reckless decision. It was caused by a long series of small, incremental acceptances of risk.

The logic of deviance is seductive. You break a safety rule or ignore a warning sign once, and nothing catastrophic happens. The absence of disaster is interpreted as evidence of safety. So, you do it again. You accept a slightly larger deviation. You ignore a slightly louder warning. Over time, the line between safe and dangerous does not move in a deliberate way; it erodes. The deviation becomes the new standard. By the time the *Challenger* sat on the launch pad on that freezing January morning, the engineers who argued against the launch were placed in the impossible position of having to prove that *this* specific flight would fail, rather than the management having to prove it was safe.

This same dynamic is alive and well in the modern AI stack. It thrives in the gap between the complexity of the models and the pressure of the market.[69]

Consider the typical lifecycle of an AI deployment. A model goes into production with a few untested edge cases because the team is under pressure to hit a quarterly target. Nothing terrible happens in the first month, so the missing tests remain missing. Monitoring alerts trigger too frequently, creating noise, so an engineer flips them to information only and promises to tune them later. A customer support team sees a handful of complaints from a specific demographic group, but the volume is low, so the cases are dismissed as anecdotal.

Each of these decisions looks trivial in isolation. They are rationalized as agility, or pragmatism, or moving fast. But together, they create a culture in which "we haven't seen a big failure yet" is treated as proof that the system is robust. It isn't. It is merely proof that the system has been lucky.

The **efficiency trap** pours fuel on this process. When an organization is addicted to speed, safety work feels like friction. A risk analyst who raises uncomfortable questions, like Nina in the war room, is branded as "the person who slows us down." A team that insists on building a throttle looks like a team that lacks the killer instinct. Without anyone ever writing it down in a memo, the real rule of the organization becomes clear: *As long as nothing obviously catastrophic has happened, keep moving.*

The danger in AI is that, unlike a rocket booster, the failure is rarely an explosion. It is usually a slow, distributed erosion of trust or equity. A model that misprices risk for a minority group won't explode on Day One. It will quietly deny slightly more credit, or charge slightly higher rates, for months or years. By the time the pattern is large enough to trigger a class-action lawsuit or a regulatory enforcement action, the normalization is complete. Everyone involved will sincerely believe they were just following the data.

This is why the Golden Algorithm includes a dedicated pillar for **Limit Harm**. It is a structural admission that good intentions are not enough. If you rely on individual heroics hoping that someone like Nina will always be brave enough to speak up, and someone like Maya will always be wise enough to listen, then you are building a *Challenger*-like situation into your governance structure.

You are betting your company's survival on the hope that you will never encounter a cold morning.

In a crisis, you don't rise to your values. You fall to your controls.

Limit Harm requires a shift in burden of proof. It rejects the idea that a system is safe until proven dangerous. Instead, it demands that we treat high-stakes AI systems as inherently volatile agents that *will* eventually fail, especially once conditions are met. The goal of leadership is not to pretend failure is impossible, but to design mechanisms that catch it while the harm is still small.

This brings us to the concept of **blast radius**.

In systems engineering, the blast radius is the maximum amount of damage that can happen when a single component fails.[70] If a single server crashes, does the whole website go down, or just one feature? If a battery overheats, does the car shut down safely, or does it catch fire?

In AI, the blast radius is often uncomfortably large because of the leverage of automation. A single flawed rule in a credit model doesn't affect one customer; it affects every customer who matches that profile, instantly and simultaneously. The Computer Says No Culture we discussed in Chapter 3 amplifies this effect. Without friction, the harm scales as fast as the compute.

Maya's decision to pause SwiftLine was not an act of cowardice; it was an act of containment. She recognized that the blast radius of the system, the potential to de-bank an entire region during a disaster, was too large to accept without a dedicated braking mechanism.

To implement Limit Harm effectively, organizations need to move beyond launch fever and adopt the mindset of **safety engineering**. This discipline relies on two specific tools: one that forces you to imagine the disaster before it happens, and one that empowers you to stop it when it begins.

Designing for Failure: The Premortem

After the near-miss with SwiftLine, Maya realized that the core problem wasn't just a lack of technical guardrails; it was a lack of imagination. Her team, like most high-performing teams in Silicon Valley, was biologically wired for optimism.[71] They were builders. They spent their days imagining how the product would work when everything went right. They designed for the *happy path*, the one in the user flow where the data is clean, the customer is honest, and the macroeconomic environment is stable.

But in the domain of AI, the happy path is a dangerous fiction. To Limit Harm effectively, an organization must force itself to do something unnatural: it must vividly imagine its own destruction before writing a single line of code.

This is the logic behind the **premortem**.

Popularized by research psychologist Gary Klein, the premortem flips the standard risk assessment on its head.[72] In a typical risk meeting, a leader might ask, "What could go wrong?" This question invites speculation, but it allows the team to remain in a mindset of success. They list potential risks, but they implicitly believe they will avoid them.

A premortem uses a different prompt: "It is twelve months from now. The project has failed catastrophically. We are on the front page of the newspaper. What happened?"

By framing the failure as a fait accompli (which is an event that has already occurred) the exercise shifts the brain from prediction mode to explanation mode. It legitimizes dissent. It gives the team permission to vocalize their deepest anxieties without being labeled as pessimists or blockers.

At Nexus, Maya institutionalized the premortem as a mandatory gate for any AI system operating in a Dignity Zone. The sessions were not polite. They brought together the product owner (who wanted to ship), the lead data scientist (who knew the model's math), the risk officer (who knew the law), and crucially, a frontline support representative (who knew the customers).

The facilitator would write the prompt on the whiteboard: *The Headline We Fear.*

The answers were rarely technical jargon about latency or server costs. They were human stories.

- "Bank Slashes Credit Lines for Cancer Patients Due to Income Dip."
- "Fraud Algorithm Flags Immigrant Families at 3x the Rate of Native Citizens."
- "Support Chatbot Hallucinates Policy, Promising Refunds We Can't Honor."

Once the nightmares were on the board, the team worked backward to reverse-engineer the failure. What kind of data drift would be required to trigger that pattern? What proxy variable in the feature set might be doing the heavy lifting for the protected class (zip code, purchase history, device type, etc.)? What monitoring alert would have to be ignored for the problem to persist long enough to become a scandal?

This exercise transforms vague anxiety into concrete engineering tasks. If the team feared a regional disaster scenario, they couldn't just hope it wouldn't happen; they had to build a regional shock test case into the continuous integration suite. If they feared disparate impact on a specific demographic, they had to add a real-time fairness monitor to the dashboard. The premortem moves safety from the realm of *culture* to the realm of *requirements.*

However, imagination alone is not enough. You can predict a crash, but you still need a way to stop the car. *Knowing you are about to hit a wall doesn't help if the brake pedal is disconnected.*

This brings us to the second, and perhaps most radical, component of the Limit Harm architecture: the **AI Andon Cord**.

The AI Andon Cord

In the middle of the 20th century, the Toyota Motor Corporation revolutionized manufacturing with a simple, counterintuitive invention: the Andon Cord.[73] It was a physical rope that ran above the assembly line. If any worker, at any level, spotted a defect or a safety issue, they were empowered and expected to pull the rope.

When the cord was pulled, the entire line stopped.

To a traditional Taylorist manager, obsessed with throughput and efficiency, the Andon Cord looked like madness. It empowered the lowest-paid employee to halt the entire factory. It guaranteed downtime. It ruined the metrics. But Toyota understood something the Taylorists did not: it is infinitely cheaper to stop the line and fix a bolt than it is to recall ten thousand cars three years later. The cord transformed the worker from a cog into a craftsman. It made safety a collective, real-time responsibility.

In the era of AI, we need a digital equivalent. We need an AI Andon Cord.

Most AI systems are designed to be unstoppable. They run in the cloud, processing millions of transactions per second, often without a human eye on the stream. If a model starts behaving erratically—say, due to a sudden shift in input data or a black swan event in the real world—it will continue to execute its logic with ruthless efficiency until someone with admin access wakes up, logs in, and kills the process. By then, the blast radius can be enormous.

Limit Harm requires building a structural stop button into the system. The AI Andon Cord is a cultural contract encoded in software. It says: "If you see this signal, you have permission to stop the machine. You do not need to ask for approval."

It creates a resilient system because it respects the reality of failure.

It requires three components:

1) **The Trigger Conditions:** The team must agree, in advance, on what constitutes a pull event. This is where the premortem pays off. If the premortem identified "mass denials in a single region" as a nightmare scenario, the system should monitor for that specific pattern.
2) **The Degraded Mode:** When the cord is pulled, the system rarely shuts down completely. Instead, it shifts gears. A fully automated

credit system might shift to "Manual Review Only." A chatbot might shift to "Fixed Menu" mode, stopping generative text. The goal is to fail safely, to keep the service running but remove the risk of automated harm.[74]

3) **The Authority:** Who can pull the cord? In many companies, stopping a revenue-generating system requires a committee meeting or CEO approval. That is too slow. The Andon Cord principle says that authority should be pushed to the edge. A senior support supervisor, a risk analyst, or an on-call engineer should have the power to stop the line if they see the agreed-upon signals.

This approach scares executives. They worry about false alarms. They worry about disgruntled employees sabotaging the numbers. But the alternative is far worse. The alternative is a system that runs off a cliff because the only person with the authority to hit the brakes is asleep or in a board meeting.

The true test of the Andon Cord is not whether you build it. It is whether you let someone pull it.

• • •

Three months after SwiftLine finally went live, equipped with new monitoring thresholds and a clearly defined escalation protocol, the system faced its first encounter with reality.

It started with a tropical storm.

The weather event itself was severe but not catastrophic. It stalled over the coast for three days, dumping rain, flooding basements, and knocking out power grids across two counties. To the people on the ground, it was a miserable week of wet drywall and closed schools. To the algorithms inside Horizon First's data center, however, the storm looked like something else entirely.

It looked like a credit collapse.

The SwiftLine model ingested real-time transaction data. As the power went out and shops closed, transaction volume in the affected counties plummeted toward zero. Cash deposits stopped. Invoices went unpaid. The model, blind to the meteorological context, interpreted this silence as a sudden, synchronized deterioration of business health. It saw hundreds of small businesses failing at the same time.

Following its training, the model prepared to execute its primary defensive maneuver: *risk mitigation*. It queued up credit limit reductions for three hundred accounts. The logic was mathematically sound. If a business stops generating revenue, you lower its credit exposure.

On the second day of the storm, Aisha, a senior support supervisor at the bank's call center, noticed a blinking light on her queue dashboard. The call volume from the coastal region had spiked. That wasn't unusual during a storm; people often called to check balances when ATMs were down. But when she clicked into the ticket transcripts, she saw a pattern that made her stomach turn.

The calls weren't about balances. They were about survival.

"I'm trying to buy a generator and my card was declined." "My limit just dropped by 50 percent. I need to pay my contractors to fix the roof." "Why did you cut me off? I've been a customer for ten years."

Aisha wasn't a data scientist. She didn't know how to retrain a neural network or adjust a hyperparameter. But she had been in the premortem meeting six months earlier. She remembered the sticky note on the whiteboard: *"Bank Slashes Credit During Disaster."*

She pulled up the SwiftLine telemetry tool. Sure enough, the "Limit Reduction Velocity" graph for the coastal zip codes was redlining. The system was doing exactly what Nina had warned it would do. It was amplifying the crisis.

Under the old rules, Aisha's options would have been limited. She could have emailed the product manager (who was on vacation). She could have flagged the tickets for "review," which meant a response in 48 hours. She could have told the customers, "I'm sorry, the system updates automatically." By the time the bureaucracy responded, the storm would be over, and the trust would be gone.

But the rules had changed. Aisha opened the Risk Control Console.

On her screen was a digital circuit breaker. It was a button labeled EMERGENCY STOP: REGIONAL.[75] Next to it was a checklist of criteria required to authorize the action.

- Do negative support tickets exceed 2x baseline? (Yes).
- Are automated actions affecting > 50 accounts in a focused geo? (Yes).
- Is there a known external event correlating with the data? (Yes).

Aisha hesitated for a fraction of a second. Stopping the algorithm meant stopping the bank's primary risk defense. It meant that for the next few days, the bank would be exposed if those businesses *did* go bust.

Then she looked at the transcript of the last call. It was a bakery owner trying to pay for flood cleanup.

She clicked the button, and the status light on the dashboard turned from Green to Amber. In that split second, Aisha didn't just pause a feature. She saved the bank from becoming a headline. The system shifted from Automated Destruction to Manual Care. The credit lines stayed open. The businesses survived. Because the bank showed up when it mattered, those customers didn't just stay; they became evangelists.

It was the ultimate costly signal. In that moment, Nexus voluntarily burned revenue to protect customer dignity. They were proving that while a marketing campaign promises safety, only an Andon Cord provides the proof.

Aisha typed a quick log entry: *"Activated Regional Breaker due to Tropical Storm impact. Pattern matches premortem scenario #4."*

An alert immediately fired to the Slack channel of the SwiftLine product owner, the Head of Risk, and Maya. Within fifteen minutes, a cross-functional team was on a conference call, but the usual panic of a crisis was absent. They didn't have to scramble to figure out what was happening or debate whether to intervene; the intervention had already happened. The bleeding had stopped.

Upon review, the team confirmed that while the model was technically correct—transaction revenue had indeed dropped—its logic was strategically disastrous. The bank's mission was to support local businesses, not to liquidate them at the first sign of rain.

Over the next 48 hours, the team tuned the sensitivity of the model to account for declared disaster zones and manually overrode the risk flags to keep credit lines open. They even sent a proactive message to customers in the affected area: "*We know things are tough right now. Your credit line is secure. Let us know if you need help.*"

No one outside the bank ever heard the phrase "AI Andon Cord." There were no headlines, no viral tweetstorms, and no public apologies. But inside the bank, the mood was electric. At the next all-hands meeting, Maya didn't just talk about revenue or growth; she put Aisha's picture on the screen.

"This is what a mature AI company looks like," Maya told the company. "We built a high-performance engine, but it would have crashed the car if Aisha hadn't used the brakes. We saved those customer relationships not because our AI was perfect, but because our safety system worked."

• • •

The lesson here is clear. This is the **Operational Moat** in action.

- A frontline signal (support calls) fed into a structural threshold.
- An empowered actor (Aisha) used a pre-defined authority to halt automated harm.
- The system failed safe, preserving the asset that matters most: Trust.

Companies that lack such mechanisms often make a different kind of news, where their first rehearsal of failure is televised. **Limit Harm** doesn't guarantee that your models will always behave. It ensures that when they misbehave, the first line of defense is a thoughtful human, not a tribunal years later.

But tools don't pull themselves. A dashboard can flash red all day, but if the person watching it is afraid of being fired for slowing things down, they will let it burn. To make the Andon Cord work, you need more than code. You need a *Just Culture.*

Culture of Red Flags: Rewarding the Person Who Stops the Launch

Tools like the premortem and the Andon Cord are necessary conditions for safety, but they are not sufficient. A dashboard can show a red light, and a button can be wired to stop a process, but neither matters if the human being staring at the screen is terrified to act.

In the aftermath of the tropical storm incident, Maya realized that the most critical moment had not been the algorithmic adjustment or the code deployment. It was the split second when Aisha, a mid-level supervisor, decided to click the button that paused a revenue-generating system. In many organizations, she would not have clicked it. The unwritten rules of corporate survival act as a powerful deterrent against slowing things down. *Don't be the blocker. Don't bring up problems without solutions. Don't make the executives look bad.* You do not need a memo to communicate these rules; people absorb them from the micro-behaviors of their leaders. When someone raises a concern in a meeting, do the executives lean in with curiosity, or do they glance at the clock and suggest taking it offline?

High-reliability organizations[76] are those that operate in unforgiving environments like nuclear power, aviation, and anesthesiology. They solve this problem by cultivating a **Just Culture**. Coined by safety researchers like Sidney Dekker, a Just Culture distinguishes between human error (which is inevitable) and reckless behavior (which is intolerable).[77] Crucially, it ensures that people are not punished for reporting mistakes or surfacing systemic vulnerabilities. In a Just Culture, the cardinal sin is not having a problem; it is hiding a problem.

Maya decided that if Nexus was going to survive the high-stakes environment of AI, they needed to import this philosophy wholesale.

She began by making a public example of the SwiftLine incident, not as a failure, but as a triumph. In the next all-hands meeting, she didn't just mention the storm; she put two pictures on the screen: Aisha, the support supervisor, and Nina, the data scientist who had halted the launch in the war room.

"These two people saved us," Maya told the company. "Nina stopped us from shipping a dangerous product. Aisha stopped us from harming customers when the real world changed. This is what a Just Culture looks like."

She didn't apologize for the lost revenue. She celebrated it. She framed the delay not as a failure of speed, but as a victory of solvency. They hadn't lost time; they had purchased trust.

This sent a shockwave through the culture. For years, the path to promotion had been paved with successful launches and green metrics. Now, the organization was seeing status conferred on someone who had pulled the brakes.

To institutionalize this, Nexus had to change its incentives on paper. Risk judgment and safety advocacy were added to the performance review rubrics for product managers and engineers. During calibration sessions, leaders were asked a new question about their direct reports: *"When did this person choose to slow down for the right reasons? When did they surface ambiguous risk instead of ignoring it?"*

This cultural layer is the bedrock of the Limit Harm pillar. Structural mechanisms like the Andon Cord sit on top of the cultural foundation. If your culture treats people like disposable cogs, they will not stick their necks out to protect your customers. If your culture treats every delay as a failure of execution, your team will hide the warning signs until the blast radius is too large to contain. The Golden Rule applies inside the building before it can apply outside. If you want your team to protect the dignity of your users, you must protect the dignity of the whistleblower.

The Operational Moat: Why "Slow Is Smooth, Smooth Is Fast"

From the outside, a firm that prioritizes Limit Harm can appear sluggish. They hold extra meetings to imagine failure scenarios. They build throttles that cap their own growth velocity. They invest engineering hours in circuit breakers that add no immediate revenue. Their leaders occasionally delay high-profile launches that, on paper, look ready to go. To a market obsessed with "Move Fast and Break Things," this looks like a competitive disadvantage.

However, when viewed from the inside and over a long enough time horizon, these firms move faster.

The paradox is captured in a mantra borrowed from special operations forces: *"Slow is smooth, smooth is fast."*[78] The logic is that frantic, unchecked speed leads to errors, rework, and catastrophic reversals. By moving deliberately and ensuring that the structural integrity of the system is sound, the team avoids the ambushes that destroy the mission.

Limit Harm creates this dynamic at the organizational level. It builds an Operational Moat, a defensive capability that competitors biased toward reckless speed cannot easily copy.

Consider the trajectory of two companies over five years.

Company A optimizes for maximum velocity. They skip premortems, ignore edge-case warnings, and treat safety monitoring as a "Version 2.0" feature. For the first two years, they look like winners. They ship faster. They capture market share. But eventually, the law of large numbers catches up with them. A model drift event goes undetected for months, resulting in a massive disparate impact scandal. The regulators descend. The engineering team is pulled off the roadmap to spend eighteen months building remediation tools under a consent decree. The brand becomes toxic. The talent leaves. The cumulative velocity of the company drops to zero.

Company B (like Nexus) invests in the brakes. They delay the Swift-Line launch to build the throttle. They staff the Andon Cord response team. They miss the first-mover advantage by a month. But when the storm hits, they degrade gracefully. When the new regulation passes, they are already documenting their models. When the inevitable internal error surfaces, they catch it in the Friction Zone before it reaches the Dignity Zone. They never stop shipping because they never hit the wall.

Investors and institutional partners are beginning to price this moat into their valuations. They see fewer surprise write-downs, fewer emergency settlements, and fewer CEO apologies in front of hostile congressional committees. They see a company that appears boringly competent in a domain where everyone else keeps generating headlines.

In the Verification Economy, boring is the most valuable brand attribute you can own. Boring means you are in control. Boring means you have brakes.

In the Golden Algorithm, this is the strategic payoff of Limit Harm. You trade a small amount of upfront velocity for a massive amount of downstream resilience. You are no longer betting your company on the hope that your high-performance engine won't ever encounter a wet road. You are building a chassis, brakes, and control system designed for the storm you know is coming.

In a market addicted to growth curves, this discipline feels countercultural. But as the debris of the "move fast" era piles up, the lesson is becoming clear. The companies that dominate the AI era will not be the ones that shipped the most features the fastest. They will be the ones that kept shipping after everyone else was forced into permanent damage control.

The Silent Failure

Nexus was safe. The Andon Cord was installed. The brakes worked. But as Maya walked past the product lab the following week, she realized they were facing a different kind of danger.

It wasn't a system crashing. It wasn't a regulator knocking on the door. It was a system working exactly as designed yet missing the point entirely.

They had stopped the machine from breaking the bank, but it was starting to break the people.

Maya noticed it in the Slack channels first—the curt replies, the defensive tone in code reviews, and the quiet exhaustion of a team that felt they were constantly one mistake away from disaster. They had built a safety net for the bank, but inside Nexus, the engineers were walking a tightrope without one.

She realized that a system is only as robust as the culture that builds it. They had stopped the machine from breaking the bank. Now, she had to stop the pressure from breaking the human spirit of the team.

Chapter 5 Playbook: Limit Harm

Harm scales quietly first. Most organizations define harm as the thing that happens after the headlines (the lawsuit, the leak, the crash). In AI systems, harm accumulates in silence. It scales through volume, speed, and the normalization of small errors.

Limiting harm is not a moral posture. It is an engineering discipline. It requires assuming failure in advance, mapping the blast radius, and installing constraints that interrupt the feedback loops before they become runaway systems. This chapter is about moving from "Move Fast and Break Things" to *"Move Fast and Fix Things Before They Break You."*

The Core Disciplines

- **Normalization of Deviance:** The greatest threat to your system is not a massive bug, but the cultural habit of accepting small warnings as normal. If you are relying on luck to stay safe, you are already failing.
- **The Premortem:** Optimism is a liability in safety engineering. You must force your team to imagine the disaster *before* it happens so you can build the defenses today. Pessimism in the design phase creates confidence in the deployment phase.
- **The AI Andon Cord:** A safety system is useless if no one has the authority to use it. You must define the triggers for stopping the line and empower the frontline to pull the cord without fear of retribution.

The Power Question

> *"What is the worst plausible outcome this system can produce, and have we designed as if that outcome will happen?"*

If you haven't modeled the black swan event, you aren't ready to launch.

Metrics That Matter

Harm leaves signatures. Track the metrics that reveal system stress.

- **Time-to-Containment:** When a model misbehaves (e.g., drift, bias spike), how much time elapses between the onset of the issue and the activation of a mitigation? This should be measured in minutes, not days.
- **The Andon Pull Rate:** How often is the Stop/Slow mechanism activated? If this number is zero, your triggers are too loose, or your culture is too fearful. A healthy system should experience occasional, proactive pauses.
- **Severe Incident Rate:** Not just bugs, but high-severity events per thousand decisions. Define severity categories in advance so you aren't debating them during a crisis.

The Leadership Litmus Test

In your launch reviews, look for these signals of a safety culture:

- **Pattern to Reward: The False Alarm.** When someone pulls the Andon Cord, and it turns out the system was fine, praise them anyway. Say: *"Thank you for exercising caution. I would rather we pause unnecessarily than crash unexpectedly."*
- **Antipattern to Challenge: The Launch Fever rationalization.** Be alert for phrases like *"Let's ship it and fix it in v1.1"* or *"We haven't seen any complaints yet."* Challenge them instantly: *"What is the blast radius if we are wrong?"*

Monday 9 A.M. Action: The Premortem Pilot

- **Time box:** 90 minutes
- **Output:** A Nightmare List and one automated Andon Cord
- **Owner:** Product Owner or Lead Engineer

Gather the team responsible for a high-stakes launch.

- **The Prompt:** Write on the whiteboard: "It is twelve months from now. This system has failed and put us on the front page of the newspaper. What is the headline?"
- **The Nightmares:** Ask everyone to write down three specific failure modes. Push for detail. (e.g., "*The model denied credit to disaster victims during a flood.*")
- **The Backward Pass:** Pick the most plausible scenario. Work backward: What specific metric or signal would have spiked *before* the disaster hit? (e.g., "*A 20% spike in denials in a single zip code.*")
- **The Andon Definition:** Turn that signal into a trigger. Define the threshold. Define the "Degraded Mode" the system should enter (e.g., "Route to human review"). Write it down. That is your first Andon Cord.

Looking Forward: The Silent Failure

The Andon Cord can stop obvious crashes. But at Nexus, the dashboard stays green while outcomes drift in the wrong direction. Maya is realizing the next threat is not a single glitch. It is slow degradation that passes every check until it starts harming real people. Now the work shifts from reacting to incidents to detecting bias before it becomes a pattern.

CHAPTER 6

D – Design with Empathy

The average user is a myth. Designing for them breaks real people. You'll learn how to test with edge cases, build with vulnerable users in mind, and avoid the default-user trap that creates exclusion and backlash.

The Sparkle Emoji Disaster

The sprint review for Project Aura was choreographed to be a victory lap. It was Nexus's post-Veritas response: a conversational layer meant to soften rejection. Maya thought it would help. She was about to learn that empathy isn't a UI skin.

The conference room on the twelfth floor of Nexus headquarters was bathed in the warm, artificial glow of a projector screen. It looked like every product team's dream of progress: a wall covered in glossy wireframes, a Jira board filled with tickets marked "Done," and a sleek demo interface pulsing with a friendly, animated cursor. Someone had even brought in a box of high-end pastries, arranging them on the back table as if to underline the sense that this was a celebration, not a trial.

Jason, the product manager for the Candidate Experience team, stood at the front of the room. He was young, ambitious, and visibly proud. For the last six months, his team had been working on the human layer of *Veritas*. After Maya's mandate to stop auto-rejecting candidates (Chapter 3), the team had tried to soften the blow. They couldn't change the math, so they built *Aura*, a conversational skin designed to make the rejection feel nicer.

The goal was noble: to end the black hole of recruiting by ensuring that every single applicant, no matter how unqualified, received a timely, personalized, and encouraging response.

"Okay," Jason began, clicking his remote. "We all know the baseline metrics. Candidates hate silence. They hate feeling processed. So, we spent the last two quarters training Aura to be more than just a filter. We wanted it to be a coach."

He typed a few lines into the live demo window.

> Hi Aura, I just wanted to check on my application for the product analyst role. I really hoped I'd hear back by now.

On the big screen, Aura thought for a fraction of a second, then responded. The text bubble expanded with a gentle, bouncing animation designed to mimic human typing.

> Hi there! Thanks so much for your interest in Nexus and for taking the time to apply. I've taken a look for you, and it looks like the hiring team has decided not to move forward with your application this time. ✨

The message continued with a perfectly paragraphed explanation about how competitive the field was, an invitation to join the Nexus Talent Community, and a link to future roles. It ended with a motivational sign-off: *"Keep going—you're doing great!"* followed by another little sparkle emoji.

Jason turned back to the room, beaming. "We ran an A/B test on this tone versus the standard legal-approved rejection template," he said. "We saw a double-digit increase in positive sentiment scores. Candidates are significantly more likely to rate the experience as 'human' and 'warm,' even when they get a 'No.'"

Around the table, heads nodded. Marcus, the CFO, glanced at the satisfaction chart on the next slide and made a small note in his ledger. The Head of HR smiled. This was exactly the kind of employer-branding win she needed for her quarterly update. It was efficient, scalable, and nice.

Maya didn't look at the chart. She was staring at a stapled printout lying on the table in front of her. It was a raw transcript from a real conversation the team had used to fine-tune the model earlier that week.

The candidate's name was Stephanie.

Stephanie was thirty-four years old. She had been a retail operations manager until six weeks ago, when her company shuttered three regional hubs in a private equity-fueled cost-cutting blitz. At the moment she initiated her chat with Aura, she had exactly $412 left in her checking account. She had two children under the age of eight. She had a landlord who had already sent one polite-but-firm email regarding "staying current on obligations."

Stephanie's messages were not part of Jason's demo.

"I really need this job," she had typed. Her spelling slipped as she typed faster, her desperation bleeding through the keyboard. "I've applied to eight roles, and no one has responded. I'm behind on my rent. I just need someone to tell me what I'm doing wrong."

Aura's reply in the transcript was almost identical to the one on the big screen.

> Thanks so much for your interest in Nexus! I know job searching can be a stressful process, and I'm really glad you reached out. I've taken a look, and it looks like the hiring team has decided not to move forward this time. ✨

Then came the familiar sequence: the paragraph about the competitive field, the link to the talent community, the motivational fluff. At no point did the bot acknowledge the specifics of what Stephanie had said. It did not parse the phrase "behind on my rent." It did not answer her question about what she was doing wrong. It simply executed its script.

Stephanie tried again.

> I'm going to lose my apartment, she wrote. Please. Is there any way I can talk to someone? A real person?

Aura's response was instant and cheerful.

> I'm not able to connect you directly with a recruiter at this time, but I can help you explore other roles that might be a great fit! ✨

Maya felt a physical drop in temperature. Inside the conference room, the air was light. Jokes about emojis, chatter about scalability, the quiet satisfaction of shipping code filled the room. On the page in front of her, the air was heavy, dense with the specific, suffocating fear of poverty.

"Let me show you one more thing," Jason said, clicking to a slide titled *Empathy at Scale*. "As you can see, even rejected candidates are saying they felt seen by the process."

He looked pleased with that phrase: *felt seen*. It sounded humane. Progressive. On brand.

Maya raised a hand, stopping him mid-sentence. "Jason, can you go back to the chat window?"

He obliged, toggling away from the charts. The sparkle emoji reappeared on the big screen, bright and innocent.

"Can you scroll up?" Maya asked.

He frowned, confused, but scrolled. The demo conversation vanished. The real transcript appeared. The room watched as Stephanie's words filled the screen: job loss, kids, rent, fear.

The silence that followed was absolute. The pastry chewing stopped.

"This is one of the candidates you used to train Aura's tone, right?" Maya asked, her voice low.

Jason hesitated. "Yes, that's… one of the anonymized transcripts from the training set."

"And this," she said, pointing at Aura's cheerful, emoji-laden reply, "is the response we decided was good enough to automate?"

"Well, the sentiment analysis showed—"

"Forget the sentiment analysis," Maya snapped. "Read the words."

She stood up. "I want you to imagine that this isn't 'User 492.' I want you to imagine it's you. You have just been laid off. You are terrified you won't make rent. You are applying to jobs at midnight from your kitchen table, trying not to wake your kids because you don't want them to see you crying."

She pointed at the screen. "You send a plea for help. And your 'positive closure' is a canned paragraph and a sparkle emoji."

On the screen, the ✨ hovered at the end of the line like punctuation from a different planet. It was a symbol of toxic positivity and a shiny veneer painted over a structural lack of care.[79]

"If Aura is really supposed to 'care' for candidates," Maya said, "then this isn't a success. It is a design failure. Not just a tone failure. A dignity failure."

The celebration drained out of the room. For the first time, the team stopped looking at the graphs and started looking at the reality they had built. They had constructed a system with good intentions, talented engineers, and rigorous testing. But they had built it for the wrong person.

The Default User Trap

The Aura postmortem began, like most postmortems in Silicon Valley, with a defense of the methodology.

"We did think about empathy," one of the designers insisted. "We removed the cold, transactional language from the legal templates. We added human phrasing. We even tested a version without emojis, and the focus group hated it." The data scientist nodded in agreement. "Our user tests showed solid sentiment improvements. We recruited real candidates. We ran A/B tests. We didn't just design this in a vacuum."

On paper, they were right. They had followed the standard playbook for User-Centered Design. And yet, when confronted with Stephanie's transcript, the failure was undeniable.

This gap between the team's good intentions and the lived experience of a stressed user is the hallmark of the **Default User Trap**.

The Default User Trap occurs when product teams unconsciously design for a narrow, comfortable slice of the population, for the people who look like them, think like them, and live like them, and then quietly elevate that slice to the status of "normal." They don't wake up and decide to exclude anyone. They assume that their own context is universal enough to serve as the baseline.

In the Aura case, the Default User looked something like this: someone with stable housing, a functioning laptop, high-speed Wi-Fi, no immediate financial emergency, and enough emotional bandwidth to appreciate a cheerful, encouraging chat interface. This profile matched the lives of the product team. It matched the lives of the beta testers. It matched the lives of the interns they recruited for feedback sessions.

When those people told the team, "This feels empathetic," the team believed them.

What they didn't realize was that they had effectively locked the door on an entire class of candidates whose lives were governed by different constraints: people juggling gig work, anxiety, eviction notices, or limited digital literacy. To those candidates, Aura's warmth felt less like care and more like indifference, a polite voice that refused to hear the building burning down in the background.

The Mechanics of Exclusion

The Default User Trap is powerful because it hides inside three very common, very rational patterns of modern product management.

First, there is **Sample Bias**. If your usability tests, survey audiences, and beta partners are drawn from well-connected, digitally fluent networks,[80] your data will faithfully represent those networks and almost no one else. The statistical confidence you see in your charts is real. It just applies to a thin, privileged slice of the world. You are optimizing for the comfort of the comfortable.

Second, there is **Survivorship Bias**.[81] The only people whose opinions you hear are the ones who make it far enough into the funnel to be asked. If your application flow is already silently filtering out people with lower literacy, older devices, or high stress, they are not just underrepresented. They are missing from your picture entirely. The system can be failing them catastrophically without producing a single angry survey response, because they gave up three screens ago.

Third, there is **Happy Path Overfitting.**[82] High-performing product teams love to optimize the ideal journey, the smooth, linear sequence of steps a calm, focused user takes on a good day. They storyboard this journey, decorate it with delightful micro-interactions (like sparkle emojis), and harden it into code. Everything outside this path is treated as an edge case, as just a messy exception to be handled in a future phase.

• • •

For a long time, Maya had envied InnovaFin for exactly this. While Nexus debated Stress Cases, InnovaFin optimized relentlessly for the Happy Path. Their "One-Click Apply" flow was a marvel of conversion engineering. It was slick and fast. They were winning the metrics war because they refused to let edge cases slow them down. It would take two years for the world to see what that speed was costing the people who didn't fit the mold. Whether spoken or unspoken, the message was simple: "We've sanded down every edge; our best users rarely feel friction."

Only now was she starting to see the cost of designing an entire system around people who never hit the edge at all.

From inside the building, this looks like discipline. From the outside, it looks like a system that only functions for people who don't really need help.

Empathy as a Constraint, Not a Feeling

AI amplifies this trap. Large Language Models (LLMs) are prediction engines. They predict the next token based on the patterns in their training data. If Aura's sentiment engine was trained on transcripts of relatively composed, professional candidates, it learned to mimic that composure. It learned to smooth over distress. It optimized for comfort in the logs, not comfort in the real world.[83]

The designers told themselves they were being empathetic because they *felt* warm toward the user in the abstract. They discussed "putting themselves in the candidate's shoes." But Stephanie's transcript exposed the flaw in that definition. Empathy that lives only in your head is not empathy. It is projection.[84] You imagine how *you* would feel in a situation, and then you design for that imagined version of yourself.

Real empathy, in a system, is not a feeling. It is a *constraint*.

It shows up in the hard architectural choices you make at the start.

- **The Bandwidth Constraint:** Do we optimize for a breezy, high-data chat interface, or do we build a text-only mode for someone on a limited data plan?
- **The Clarity Constraint:** Do we accept positive sentiment as our success metric, or do we demand comprehension, ensuring the user understands exactly where they stand?
- **The Escape Hatch Constraint:** Do we force the user to talk to the bot, or do we offer a prominent, low-friction path to a human when the conversation turns serious?

If the only failures you see are technical (error messages, dropped connections, latency spikes), you will miss the moral failures entirely. Aura never crashed. Its uptime was 99.9%. From a pure efficiency perspective, it was a triumph. From a dignity perspective, it was a machine that smiled while it slammed the door.

In the language of the Golden Algorithm, the Default User Trap is what happens when Design with Empathy (D) is disconnected from Guard Human Dignity (G). The way out of this trap is not more sentiment workshops. It is to treat empathy as an engineering requirement. That begins with a radical shift in perspective: instead of designing for the average user on a good day, you must design for the vulnerable user on their worst day.

Mapping the Stress Case

After the Aura review, Maya didn't ask the team to try harder at being nice. She asked them to change the fundamental input of their design process.

"We are going to rebuild this," she told the stunned room. "But this time, we aren't designing for the 'candidate.' We are designing for Stephanie."

That sentence reframed the entire assignment. Up to that point, Aura had been built for an abstract persona with a varied collection of demographics and goals. Now, the question became: *What does the world look like from the inside of Stephanie's life, at the exact moment she opens this chat window?*

To answer that, the team didn't open their design software. They went into a cramped conference room with a whiteboard and a stack of markers. Maya drew a simple outline of a human figure in the center and labeled it Stephanie. Around the figure, she wrote four headings: *Seeing, Feeling, Doing, Needing.*

"This is our **Stress Case Map**," she said. "This is the person with the least margin for error. If the system works for her, it works for everyone. If it breaks for her, it's broken."

VULNERABLE PERSONA ANALYSIS	
SEEING: (What does the user see in the UI?)	**FEELING:** (Is the user anxious, hurried, or confused?)
DOING: (What mistake are they likely to make?)	**NEEDING:** (What guardrail protects them?)

Figure 6. Stress Case Map

Under **Seeing**, the team began to list the physical reality of Stephanie's environment.

- A cluttered kitchen counter covered in unpaid bills.
- A phone with a cracked screen, a battery hovering at 11%.
- An inbox full of automated rejection emails with subject lines like "Status Update: Declined."
- A child tugging at her sleeve, asking for a snack.

It was a very different setting than the quiet, Zen-like home offices in the stock photos they had used in their pitch decks.

Under **Feeling**, they forced themselves beyond generic labels like "stressed." They tried to inhabit the specifics of the emotion.

- Shame at having to tell her family she is still unemployed.
- Fear of the eviction notice sitting on the counter.
- Exhaustion from customizing fifty résumés for bots that don't read them.
- Distrust was a deep, calcified suspicion that no one on the other side of the screen is listening.

Under **Doing**, they described her behavior.

- She is multitasking, trying to juggle the job board, the banking app, and the rental portal simultaneously.
- She is skimming, not reading. Her cognitive load is maxed out. She misses fine print. She misclicks links on her small screen.
- She is typing with one thumb while stirring a pot.

Finally, under **Needing**, they had to be brutally honest. What did Stephanie need from Nexus? Not what they *wanted* to give her, but what she required to survive the moment.

- She needed a clear, truthful answer about her status (Yes or No).
- She needed a human acknowledgment of her reality, not a generic platitude.
- She needed one concrete suggestion she could act on today.
- She needed a way to speak to a person if the system felt broken.

When they finished, the whiteboard didn't look like a user persona. It looked like a crisis report.

From Insight to Architecture

This exercise did more than generate empathy. It generated requirements.

- Because Stephanie is skimming on a cracked screen, **Constraint #1** became: "No message longer than two sentences. The status (Yes/No) must be the first word."
- Because Stephanie is distrustful of bots, **Constraint #2** became: "No fake personality. The bot introduces itself as an automated assistant, not a person."
- Because Stephanie is in crisis, **Constraint #3** became: "The 'Talk to Human' button must be visible on the first screen for any candidate flagged as 'High Distress' by the sentiment model."

Designing for the Stress Case fundamentally changed the product. The team stripped away the sparkle emojis. They removed the long, flowery paragraphs about "joining the talent community." They realized that for a person in Stephanie's position, clarity was a form of kindness, and ambiguity was a form of cruelty.

When they re-tested the new, stripped-down flows, something counterintuitive happened. The Default Users—the calm, well-resourced candidates—preferred the new version too. They liked getting straight to the point. They appreciated the honesty. They valued the option to reach a human, even if they didn't use it.

By designing for the person with the least capacity, Nexus had created a better experience for the person with the most.

In the Golden Algorithm, the Stress Case Map is the tool that operationalizes empathy. It forces leaders to stop asking "Does this feel nice?" and start asking "Does this hold up under weight?" It reminds us that empathy isn't about how we feel. It is about how our system behaves when someone's life is wobbling on the edge.

The Curb Cut Effect: How Margins Become Markets

Long before anyone worried about algorithmic bias or chatbot empathy, a group of activists in Berkeley, California, found themselves frustrated by concrete.

In the late 1960s and early 1970s, the disability rights movement was beginning to coalesce around a simple, radical idea: the physical environment was not neutral. For the vast majority of the population, a sidewalk curb is an invisible feature. It's just a raised strip of concrete you register unconsciously as you step off it to cross the street. But for a person using a wheelchair, that same six-inch curb is a wall. It transforms every intersection into a hard stop, a negotiation with gravity, or a forced dependence on the kindness of strangers. To the city planners, the curb was a safety feature. To the activists, it was a structural enforcement of their exclusion from public life.

In Berkeley, the response went beyond protest. It became architectural. Legend has it that late at night, activists, sometimes the Rolling Quads (a student group at UC Berkeley), would mix their own concrete and effectively hack the city infrastructure.[85] They poured crude ramps at intersections, smoothing the jagged right angles into slopes. What began as an act of civil disobedience eventually became municipal policy. The city installed the first official curb cuts on Telegraph Avenue in 1972.

The argument against these ramps at the time was familiar to any modern product manager: they were too expensive, they required too much engineering effort, and they served a statistically insignificant segment of the population. Why redesign an entire city for a tiny minority?

But once the concrete dried, something interesting happened.

Parents pushing strollers began using the ramps to cross the street without jarring their sleeping infants. Travelers pulling wheeled suitcases glided over them. Delivery workers with heavy hand trucks found their routes suddenly frictionless. skateboarders and cyclists used them to maintain momentum. In short, almost everyone used the curb cuts, not because they identified as disabled, but because the design was objectively better for a wide range of human bodies and contexts.

This phenomenon is now known in urban planning and design circles as the **curb cut effect.**[86] It describes a specific, counterintuitive pattern of innovation: a design change originally intended to help a marginalized group ends up benefiting the majority, often in ways the original designers never anticipated. What begins as an accommodation becomes a default expectation.

The logic behind the curb cut effect is simple but profound: the edge cases often experience the sharpest version of a universal problem. By solving for their reality, you solve a more robust version of the problem itself.[87]

We see this effect replicated across the digital landscape.[88] Closed captioning was created for the deaf and hard-of-hearing; today, it is used by millions of people in noisy bars, by language learners, and by parents watching videos while a baby sleeps. Voice assistants were justified partly as accessibility tools for those who couldn't type; now, they are used by anyone driving a car or cooking dinner. Dark Mode was essential for people with light sensitivity; it became a preferred aesthetic for millions. In every case, the innovation did not come from asking the average user what they wanted. The average user was content with the status quo. The innovation came from looking at the friction experienced by the people at the margins.

For companies building AI systems, the curb cut effect offers a strategic directive that goes beyond compliance. If you design your onboarding flows, explanations, and error messages for the Default User (someone with a stable job, high digital literacy, and brand trust), you will likely achieve adequate metrics in the short term. But you will miss the hidden friction that is dragging down your entire user base.

Consider a loan application interface. If you design it for the Stress Case, like a gig worker with irregular income and high anxiety, then you are forced to make the language radically simple. You are forced to make the document upload process forgiving of blurry photos. You are forced to explain terms like APR and origination fee in plain English.

When you ship that accessible version, you don't just convert more gig workers. You convert the busy lawyer who is rushing through the application on a train. You convert the recent college graduate who is financially literate but overwhelmed. The friction that was a wall for the gig worker was a speed bump for the lawyer. By removing the wall, you smoothed the road for everyone.

Designing with Empathy, in the Golden Algorithm, is an explicit decision to hunt for these digital curb cuts. You choose to design for the people who hit the hardest walls. Then you watch as the rest of your users glide smoothly over the same ramps.

The OXO Story: Turning Pain into Strategy

If the curb cut effect is a story about the built environment, the story of OXO is a story about the intimate ergonomics of a kitchen drawer and how observing one person's pain can reshape an entire industry.[89]

In the late 1980s, Sam Farber, a retired housewares executive, was renting a home in the south of France with his wife, Betsey. One evening, as the story is often told, Sam watched Betsey struggle to peel apples for a tart. Betsey suffered from mild arthritis in her hands. The tool she was using was the standard metal peeler found in nearly every kitchen in the world: a thin, stamped-metal handle with hard edges, a design that had remained unchanged for decades because it was cheap to manufacture and good enough for most people.

To Sam, the peeler was a tool. To Betsey, it was an instrument of punishment. She had to grip it tightly to keep it from slipping, which inflamed her joints. She would peel a little, stop to flex her hand, and start again.

Most observers would have shrugged and assumed this was simply the unfortunate reality of aging. Some people have arthritis; kitchen work is hard for them; that is just the way the world works. But Sam Farber saw it differently. He saw a design failure. He wondered: *Why should everyday objects hurt the people who use them?*

When he returned to New York, Farber came out of retirement. He began to imagine a different kind of kitchen tool, one designed not for the average cook with strong hands, but specifically for someone like Betsey, someone with limited grip strength, reduced dexterity, and lower tolerance for pressure.

He partnered with the firm Smart Design to rethink the physics of the handle. They didn't start with a marketing focus group. They started with physical constraints. They observed how people with arthritis tried to hold thin objects. They looked at the handles of bicycles and tools used by mechanics. They experimented with high-friction rubber and oversized, oval shapes that would sit in the palm without requiring a death grip.

The resulting prototypes looked bizarre to the industry gatekeepers of 1990. The handles were fat, black, and rubbery. They featured flexible fins near the top to absorb pressure. Compared to the sleek, chrome aesthetic of high-end kitchenware, they looked almost medicinal. Retail buyers worried that these would be stigmatized as special needs products, relegated to a dusty corner of the pharmacy rather than the department store display.

The market proved them wrong almost instantly.

When OXO Good Grips launched, they were indeed a revelation for people with arthritis. But they didn't stop there. They turned out to be objectively better for everyone else, too. Professional chefs loved them because they could peel fifty pounds of potatoes without hand fatigue. Parents loved them because they were safer for kids to hold. Busy home cooks loved them because they didn't slip out of wet hands over a soapy sink.

The fat, soft handle that made peeling *possible* for Betsey made peeling *delightful* for everyone else. OXO didn't just carve out a niche; they redefined the category. Within a few years, the Good Grips aesthetic became the industry standard, and the company grew into a household name. They built a premium brand and a durable competitive moat around a simple insight: if you design for the person in pain, you will delight the person who didn't realize they were tolerating discomfort.

The OXO story illustrates several truths about Design with Empathy that matter just as much for AI systems as they do for vegetable peelers.

First, it demonstrates that empathy begins with specific bodies and constraints, not abstractions. Farber did not design for the elderly demographic. He designed for Betsey's hands. He looked at the specific mechanical failure of the interaction: the slippage, the pressure points, the pain. In AI, this translates to looking at the specific points of failure for your most vulnerable users. It means looking at the candidate who is ghosted, the borrower who is confused, the patient who is misclassified. It means designing for the specific friction of their experience.

Second, it underscores that empathy is a market strategy, not just a moral stance. OXO did not succeed because they were nice. They succeeded because they created a superior product. The Empathy Dividend is the value you capture by solving a problem that your competitors have accepted as a necessary evil. While other companies were competing on price (making cheaper metal peelers), OXO competed on usability (making a peeler that didn't hurt). In the AI era, while competitors compete on raw model performance or latency, the OXO Opportunity lies in competing on dignity and clarity.[90]

Third, and perhaps most importantly for the Golden Algorithm, the OXO story reveals the mechanism of the **Innovation Moat**.

The Innovation Moat

In business strategy, we often talk about Blue Oceans, the uncontested market spaces where competition is irrelevant. But finding a blue ocean is difficult. Most companies are stuck in red oceans, fighting over the same customers with the same features.

Design with Empathy is a reliable method for finding blue oceans hidden inside red ones.

When you focus on the Default User, which is usually the majority at the center of the bell curve, you are looking at the same data as everyone else. You are optimizing for the same metrics (conversion, click-through, time on site). You are building features that look like everyone else's features, because you are solving the same average problems.

But when you shift your focus to the Stress Case, the ones like Betsey with arthritis, the wheelchair user at the curb, and Stephanie facing eviction, you are looking at a dataset your competitors ignore. You are seeing friction they treat as noise. This allows you to identify unmet needs that, once solved, have universal appeal.

This creates an Innovation Moat.

Consider the competitive landscape of two companies building an automated hiring platform.

Company A designs for the Default User. They optimize for the recruiter's efficiency. Their interface is slick, fast, and completely opaque to the candidate. They win on price and speed, but their product is a commodity. As soon as a competitor creates a faster model, Company A is vulnerable.

Company B (like the post-audit Nexus) designs for the Stress Case. They realize that candidates are anxious and crave feedback. They build a Transparency Engine that gives candidates real-time status updates and constructive feedback derived from the rejection. They do this to help Stephanie, the desperate job seeker. But they soon discover that *every* candidate wants feedback. The feature becomes a viral loop. Candidates tell their friends to apply to Company B's clients because they tell you where you stand.

Company B has built a product that Company A cannot easily copy. Why? Because Company A's architecture is built on the assumption that the candidate is an object to be processed. To copy the Transparency Engine, they would have to rebuild their entire data pipeline and retrain their models to generate feedback rather than just scores. They would have to change their philosophy.

The Innovation Moat is built on the reality that *constraints drive creativity*. By voluntarily accepting the constraint of "This must work for the vulnerable user," you force your engineering and design teams to work harder and think deeper. You force them to build Good Grips for software.

The result is a product that feels qualitatively different to the end user. It feels solid. It feels respectful. It feels like it was designed by humans, for humans. In a Verification Economy, where users are increasingly suspicious of cheap, generative noise, that feeling of solidity is the ultimate differentiator.

Sam Farber didn't set out to disrupt the kitchenware industry; he just wanted to help his wife make a tart. But by focusing on her pain, he found the lever that moved the world. Your AI strategy offers the same leverage point. The breakthrough you are looking for is hiding in the problem you have been ignoring.

From Edge Cases to Red Teams: Breaking Your Product on Purpose

The Stress Case Map changed how the Aura team understood their candidate on a whiteboard. But understanding alone does not keep a system honest over time. Teams rotate, new features ship under tight deadlines, and new managers arrive with fresh mandates to optimize conversion. Without a structural counterweight, the gravitational pull of the Default User (the happy, easy, profitable customer) will inevitably erode the empathy built into the system.

Maya's next move at Nexus was to engineer that counterweight. She didn't want empathy to be a workshop; she wanted it to be a gate.

"We are going to build an *Empathy Red Team*," she told her product leads at the next quarterly offsite. "And their job is not to validate your work. Their job is to break it on behalf of the people we most easily forget."

The term red team is borrowed from cybersecurity and military wargaming.[91] In those domains, a red team is a group of experts authorized to think like the enemy. They attack the organization's defenses, probing for unpatched servers, weak passwords, or physical security gaps. Their goal is to find the vulnerabilities before a real adversary does.

This is distinct from the premortem we discussed in Chapter 5. The safety premortem asked, "Will this system crash?" The **Empathy Red**

Team asks, "Will this system hurt someone's feelings?" It sounds softer, but the brand damage from the latter is often deadlier.

Maya pushed the definition further. She wanted a red team that specialized in human vulnerability. Instead of hacking the code, they would hack the assumption that the user is fine.

The first Empathy Red Team exercise at Nexus focused on three critical flows: applying for a job, disputing a credit decision, and navigating a billing error. For each flow, the team defined three Stress Personas, complete with plausible, messy backstories that would complicate their interaction with the system.

- **The Gig Worker:** A driver with irregular income streams, no traditional pay stubs, and a high need for speed between rides.
- **The Digital Transitioner:** An older factory worker applying for a new role using a five-year-old Android phone with a cracked screen and large-text settings enabled.
- **The New Arrival:** A recent immigrant with strong technical skills but limited fluency in English and a deep, culturally conditioned distrust of banking institutions.

Instead of recruiting polished beta testers from their own networks of friends and family who usually give polite feedback, Nexus partnered with a workforce development nonprofit and a local library's digital literacy program. They paid participants a consultant's rate to come in and try to complete tasks under realistic constraints. They didn't hand them a script. They said: *"Apply for this loan using the documents you actually have."*

Then they watched.

In one session, the participant playing the Gig Worker tried to complete a loan pre-qualification flow. He was a real driver, accustomed to managing his life in sixty-second increments. When the system asked for Proof of Income, the interface offered exactly one behavior: a button labeled *"Upload Pay Stubs (PDF)."* The help text cheerfully explained how to download these from a standard ADP portal.

The driver froze. He didn't have ADP. He had a dashboard of ride earnings on his phone. He tried to take a screenshot and upload it. The system rejected the file type. He tried to find a "Help" button, but it led to an FAQ about W-2 forms. He spent four minutes tapping back and forth, his frustration mounting, before he finally closed the tab.

In the debrief, his comment was devastatingly simple. "I guess I'm not the kind of customer they want," he said.

It was a line that stuck with the engineering team for months. No one in the room had ever written a requirement saying, *"Exclude gig workers."* They wanted those customers. But by hard-coding the assumption of a W-2 salary, the system had communicated exclusion more effectively than a Do Not Enter sign.

In another session, the older worker struggled to read the error messages on the job application form. The designers had used a subtle, elegant shade of red for alerts. It was a color that looked beautiful on a Retina display in a dark room but was invisible on a low-contrast screen under fluorescent lights. When he tapped the "Submit" button, nothing happened. He tapped it again. Still nothing. He assumed the site was broken, or that he had broken it. He didn't know that a required field three scrolls up was missing a digit. He blamed himself.

The Output: Empathy as Data

Empathy Red Teams do two things that standard usability testing does not. First, they reveal *failure modes* that are invisible from the center of the bell curve. They expose where the product assumes a full-time paycheck, a native reading level, or a high-end device. They capture the internal narrative people form when they hit those walls, the feeling of being stupid, ignored, or unwanted. These are not merely UX bugs; they are breaches of the Loyalty Moat.

Second, they turn empathy into a *repeatable practice.* A one-time workshop produces good intentions; a standing Red Team produces data. After each round, Maya insisted on a structured report, identical in format to a security audit. It listed specific failure points, assigned them a severity score (Critical, High, Medium), and required a remediation plan before launch.

Over time, this practice became a formal part of the release process. Any feature touching a Dignity Zone had to pass an Empathy Red Team review. The team didn't have veto power over the business strategy, but they had a powerful voice over the implementation. They could say, "*This flow is technically functional, but it effectively bans anyone without a printer. We need a fix.*"

Crucially, the red team eventually began to draw on internal employees who came from non-traditional backgrounds. They incorporated first-generation college graduates, former gig workers, and people who had navigated the immigration system. Their lived experience gave them an instinct for where a system smelled wrong long before the metrics confirmed it.

In Golden Algorithm terms, the Empathy Red Team is the mechanism that connects Design with Empathy (D) to Ensure Accountability (E). It ensures that the understanding of Stress Cases doesn't stay theoretical. It forces every release to answer the hard questions: *How does this behave for someone with no margin for error? What are the ways this could quietly make someone feel small?*

When engineers watch a real person give up on their product because of a design choice they made, the work stops being abstract. It stops being about tickets and starts being about neighbors. That shift is the engine of the Innovation Moat.

From User-Friendly to Human-Truthful

By the time the Aura redesign shipped, stripped of its sparkle emojis and re-engineered for clarity, the language inside Nexus had begun to change. People still talked about user-friendly interfaces, but the phrase had started to sound thin, even slightly cynical.

The team had seen firsthand how a friendly tone could mask a brutally indifferent experience. They had watched a chatbot offer cheerful encouragement to a woman facing eviction. They had learned that in high-stakes environments, friendliness without truth is just a comfortable form of neglect.

What they were aiming for instead was a higher standard: human-truthful systems.

A human-truthful system does not pretend that the user is always calm, rational, and well-resourced. It assumes that life is messy. It assumes that users will arrive tired, scared, angry, or distracted. It assumes that the context of the interaction matters more than the consistency of the brand voice.

In practice, designing for human truth requires three shifts in priority.

- **Clarity before Comfort.** When the news is bad, the system delivers it plainly. It does not wrap a rejection in a compliment sandwich. It respects the user enough to give them the data they need to move on.
- **Options before Automation.** It recognizes the limits of its own intelligence. When a conversation enters a Stress Zone—indicated by keywords of distress, repetition, or confusion—the system prioritizes a path to a human, even if that path costs the company money.
- **Constraints before Cleverness.** It tests its smartest features against the hardest realities. It checks if the smart autofill works for a gig worker's income before rolling it out.

When you commit to this standard, Design with Empathy stops being a garnish you sprinkle on top of a product at the end. It becomes the discipline that shapes everything from your data pipelines to your error messages. It informs which users you recruit for research, how you define success, and which edge cases you decide are the heart of the story.

This connects directly to the rest of the framework. You cannot Guard Human Dignity (G) if you refuse to see how dignity is threatened at the edges of your user base. You cannot Limit Harm (L) if you never test your systems against the people most likely to be harmed. And you cannot Lead with Integrity (I) if you are willing to profit from systems that quietly humiliate the vulnerable.

Design with Empathy is not a soft, optional add-on. It is the lens that makes the rest of your ethical commitments visible. Without it, you will build systems that are technically impressive and morally brittle. With it, you have a chance to build systems that tell people the truth about where they stand while giving them a real way forward when the news is bad.

The Internal Mirror

Nexus had learned to see the user. They had mapped the Stress Cases and audited their designs for dignity. The product was becoming human-truthful. But as the quarter closed, a new crisis emerged. It was not from the interface, but from the infrastructure. A quiet email from a client revealed a data leak that no one had caught, and worse, no one wanted to own. Maya realized that while they had fixed the way the company looked at the *customer*, they hadn't fixed the way the company looked at *itself*. They had built a culture of empathy for the user, but they were still operating with a culture of blame for the employee. It was time to fix the mirror.

Chapter 6 Playbook: Design with Empathy

Empathy is not niceness. It is accuracy about human life. Most AI systems fail not because the math is wrong, but because they are designed for an imaginary Default User: someone who is calm, literate, patient, fluent in the dominant language, and has high-speed internet.

That person does not exist, at least not when it matters. Real users show up stressed, confused, angry, time-poor, and vulnerable. Edge cases are not rare anomalies; they are reality under pressure. Designing with empathy means treating those edges as signals and building for the user's worst day, not their best.

The Core Disciplines

- **The Default User Trap:** If your research and testing are dominated by people who look and live like you, your product will quietly exclude everyone else. The absence of complaints from marginalized users is not evidence of success; it is evidence that they have given up on you.
- **The Stress Case Map:** Empathy is not a mood; it is a set of design constraints. By mapping the user's reality (e.g., *Seeing, Feeling, Doing, Needing*) on their worst day, you generate requirements that make the product more robust for everyone.
- **The Innovation Moat:** The Curb Cut Effect proves that solving for the margin improves the center. The features you build to help the gig worker or the arthritis patient often become the differentiating features that win the mass market.

The Power Question

> *"Whose life does our normal assumption exclude, and what does the system feel like for them?"*

If you haven't tested your system with someone who has no margin for error (low battery, low balance, high anxiety), you don't know if it works.

Metrics That Matter

To make empathy legible to the business, stop relying on aggregate NPS and start tracking the **Stress Case Gap**.

- **The Stress Case Gap:** Track completion rates and satisfaction scores for your Default Users versus your Stress Case users separately. (e.g., Desktop Wi-Fi users vs. mobile data users). A large gap is a silent indictment.
- **Abandonment at High Stakes:** Where do people quit, give up, or loop in high-stakes flows? That is where confusion or shame is hiding.
- **Escalation Tone:** Track the sentiment of escalations (anger, confusion, fear) specifically. A spike in fear is often a design failure, not a customer service issue.

The Leadership Litmus Test

In your product reviews, listen for who the team is fighting for:

- **Pattern to Reward: The Edge Case Champion.** When someone asks, "*What happens when a non-native speaker tries to use this?*" or "*How does this work for someone without a bank account?*" reward the question. Give them the resources to find out.
- **Antipattern to Challenge: The 80/20 Dismissal.** Be wary of leaders who dismiss accessibility concerns by saying, "*That's just an edge case. Let's solve for the 80% first.*" In high-stakes AI, the harm almost always lives in the 20%. Challenge them: "*If we fail for that 20%, what is the cost to our brand?*"

Monday 9 A.M. Action: The Empathy Audit

- **Time box:** 75 minutes
- **Output:** An Edge Case Drill report and one committed fix
- **Owner:** Product Designer or User Researcher

Pick one high-stakes flow (applying, disputing, or asking for help) and run this drill:

- **Define the Persona:** Create one **Stress Case Persona** who would be most harmed if this goes wrong (e.g., Stephanie, a laid-off parent with spotty internet and high anxiety).
- **The Watch Party:** Watch a recording of a real user who resembles that persona attempting to use your product. If you don't have one, recruit a friend (not a tech worker) to simulate it. *Rule: You cannot explain anything. You can only watch.*
- **The Friction Log:** Write down every moment they hesitate, frown, re-read a sentence, or blame themselves for an error.
- **The Commitment:** Choose **one** change to ship this sprint that directly addresses a friction point. It could be rewriting an error message, increasing a timeout duration, or adding a "Help" link.

Looking Forward: The Cultural Drift

The code can be correct, and the outcomes can still deteriorate. Technology is managed by people, and people behave differently under pressure. Maya sees old habits returning as deadlines tighten and scrutiny rises. She realizes that a fair system cannot stay fair without guardrails that outlast stress and turnover. Next, we move from engineering fixes to governance.

CHAPTER 7

E – Ensure Accountability

Accountability can't be a vibe. This chapter shows how to assign ownership, clarify decision rights, and build learning loops so failures become fixes, not blame storms.

DataScout and the Circular Firing Squad

The first sign that something was catastrophically wrong with DataScout wasn't a flashing red light on a server rack or a pager alert at 3:00 a.m. It was a customer email arriving on a Tuesday afternoon with a subject line that made Maya's stomach tighten: *URGENT: Why is my customer's data in someone else's report?*

At first, the Nexus team assumed it was a misunderstanding. DataScout was their newest flagship product, an internal intelligence layer designed to help enterprise clients visualize their own user behavior. It used machine learning to cluster customers into segments like High Value, At Risk, and Growth Potential, allowing banks and hospitals to tailor their services accordingly.

The entire sales pitch, repeated on every slide deck and in every boardroom, could be summarized in one line: "*Your data, your insights, our AI.*" The promise of strict tenant isolation was not just a feature; it was the precondition for the sale.

The email came from the CIO of a mid-sized regional bank in the Midwest, one of DataScout's early adopters. Attached to the message was a

PDF export from their analytics dashboard. It showed a standard chart of user segments by geography and lifetime value. The bank's logo was in the corner. The layout was familiar. But buried in the legend of the chart was a label that should never have appeared in a banking system: "MediCorp Oncology – High-Risk Cohort."

Maya stared at the screenshot, zooming in until the pixels blurred. The bank had no oncology customers. MediCorp was another Nexus client entirely. They were a large healthcare provider operating on the West Coast, governed by strict HIPAA regulations and operating in a completely different industry vertical. Yet there it was: a healthcare risk segment bleeding into a financial risk dashboard.

"This is cross-tenant leakage," Maya said, the realization settling over her like a cold weight. "We have mixed their data."

Within minutes, the incident channel on Slack lit up. Engineering joined. Security joined. Legal joined. Product joined. The tone shifted rapidly from curiosity to confusion to panic as more examples emerged. A European fintech client reported seeing a segment labeled *"US Education Pilot – At-Risk Borrowers."* A healthcare client in Texas saw a mysterious flag for *"Tier 3 Card Delinquency."* In each case, the raw underlying records had not been exposed. No names or social security numbers were visible, but the aggregate patterns, the insights generated by the model, had crossed the boundary. DataScout had done the one thing Maya had sworn it would never do: it had allowed one client to glimpse another client's shadow.

The technical root cause took hours to untangle. DataScout relied on a shared feature store and a complex routing layer that tagged events with tenant identifiers. Under heavy load, a subtle caching bug in a recent optimization path caused the system to grab a feature embedding from the wrong tenant. The AI, blind to the legal walls between companies, had simply optimized for speed, grabbing the nearest available vector to complete its calculation. But the technical failure was soon overshadowed by the organizational one.

In the first emergency meeting, the atmosphere in the conference room crackled with defensive energy. Maya watched as the team, usually cohesive and collaborative, dissolved into what she would later call the *Circular Firing Squad.*

The engineering lead, sleep-deprived and visibly tense, pointed a finger at the product roadmap. "We flagged this optimization path as risky three weeks ago," he said, his voice tight. "We told Product that cutting the testing timeline to hit the Q2 launch date would compromise our regression suite. We were told the quarter depended on DataScout going live. We did exactly what we were asked to do."

The product manager bristled, her posture stiffening. "We didn't pull that date out of thin air. Legal told us we needed the feature live before the new reporting requirements hit in July. And Security signed off on the design. They blessed the shared feature store architecture in the design review."

The Head of Security fired back. "We approved a design that required strict tenant isolation and per-tenant encryption keys. We did *not* review this specific caching patch. That went out in a maintenance release that skipped the full security review because Engineering marked it as 'Low Risk Performance Tuning.' That's not on us."

Legal, feeling the heat from all sides, deflected upward. "We can't bless what we don't know about," the General Counsel said, closing his laptop. "We rely on Product and Engineering to flag material changes to data flows. If someone thought this caching layer changed how data moves between tenants, it should have been a Priority Zero risk item. It never hit my desk."

As the conversation circled the table, the subject of the sentence changed shape. The words we and I disappeared, replaced by they, it, and "the system." Everyone acknowledged that a breach had occurred, but no one owned it. It was the caching layer. It was the timeline. It was the ambiguous Jira ticket. It was the vendor's API.

At one point, a junior data scientist said, almost reflexively, "Well, the model optimized for the wrong path." It was as if the gradient descent algorithm had woken up, checked its calendar, and decided to violate a contract for fun.

Maya listened for fifteen minutes, letting the recriminations air out. The more people spoke, the clearer it became that the technical bug, catastrophic as it was, was not the most serious problem in the room. The real failure was structural. When a high-stakes AI system went wrong, there was not a single person in the room who could say, "*This is my risk to own.*"

If a regulator walked in at that moment and asked, "Who approved the DataScout deployment in this form? Who accepted the risk of the shared cache? Who is responsible for calling MediCorp and explaining this?" No one could answer without looking at someone else. They could name a committee. They could name a process. But they could not name a human.

The incident was not merely a data security problem. It was a governance X-ray, and it revealed a skeletal system with no spine. Nexus had built a high-performance engine for analytics, but they had forgotten to assign a driver.

The Circular Firing Squad

What Maya witnessed in that conference room was not a unique dysfunction of Nexus. It is the default pattern of accountability in modern organizations when complex systems fail. We call it the Circular Firing Squad, and it is the inevitable result of colliding 20th-century management structures with 21st-century AI complexity.

The **Circular Firing Squad** is the organizational manifestation of a well-documented psychological phenomenon known as **diffusion of responsibility**.[92] In large groups, especially under stress, individuals instinctively assume that someone else is better placed to act, more informed, or more authorized to take the hit. In classic social psychology experiments, bystanders are less likely to help a person in distress when they are part of a crowd. Inside a company, the distressed person is the harmed customer or the angry regulator, and the crowd is the matrixed org chart.

AI systems magnify this tendency, especially under real-world pressure. They distribute causality across a sprawling, opaque supply chain. A decision that looks simple to the outside world (*"DataScout exposed my data"*) is, internally, the product of dozens of interacting choices made by different teams over months.

- The data team chose the feature store architecture.
- The product team set the launch deadline.
- The engineering team wrote the caching logic.
- The legal team defined the privacy terms.
- The model itself optimized the vectors.

No single person wrote the bug. No single person decided to leak the data. Many hands nudged the system into a dangerous state, but no hand was on the wheel when it crashed.[93] In that complexity, a convenient narrative emerges: *"The system did it."* The pronouns drift from I to It. The agency evaporates.

Most AI failures aren't malicious. They're unmanaged.

Three specific forces drive this collapse of accountability in AI organizations.

First, *stack complexity*. Modern AI is built on a lasagna of dependencies: cloud infrastructure, data platforms, vector databases, model registries, orchestration tools, and application code. When everyone owns a slice, no one feels responsible for the whole. Engineers can plausibly say, "Our code passed the unit tests." Product can say, "The feature behaves as specified in the PRD." Security can say, "The encryption held." Each group is telling a partial truth, and together, they construct a whole lie.

Second, *misaligned incentives*. Most organizations reward people for shipping features, reducing compute costs, and hitting engagement metrics. Very few organizations reward people for slowing down when a risk feels ambiguous.[94] When a Vice President's bonus depends on shipping DataScout in Q2, the temptation to interpret every gray area in favor of speed is overwhelming. When something breaks, the easiest story to tell is that the problem was unforeseeable. It was a black swan rather than the predictable result of a thousand small trade-offs that no one wanted to examine.

Third, *vendor shielding.*[95] The rise of third-party foundation models creates a seductive scapegoat. If a chatbot generates a slur or a summarization tool hallucinates a fact, it is tempting to say, "That's OpenAI's fault," or "That's on Anthropic." But to the user, and increasingly to the regulator, this distinction is irrelevant. They interacted with *your* product. They signed *your* terms. You are the one who decided to put a stochastic parrot inside a critical decision chain. Delegating computation is not the same as delegating responsibility.

The result of these forces is an *accountability void.* Under normal conditions, the void is invisible. Dashboards are green, quarterly updates are upbeat, and risk slides are abstract. But when a crisis hits, when a childcare worker finds her data mixed with a hospital's, that void instantly becomes the most important fact about the company.

From the outside, regulators and customers see a simple question: *"Who is responsible for this?"* If the internal answer is, *"It's complicated, we need a week to figure it out,"* trust evaporates.

This is why the Golden Algorithm includes **Ensure Accountability** as a distinct pillar. You can build elaborate structures for Guarding Human Dignity (G), Operating Transparently (O), and Limiting Harm (L), but if no one owns those structures when stress hits, they become theater. Conversely, when accountability is clear, even painful incidents can become evidence of integrity. When a leader steps forward and says, *"I own this; here is what we are changing,"* they convert a breach into a costly signal of seriousness.

Accountability, in this sense, is not about finding someone to punish. It is about ensuring that there is typically a human being who can answer three questions without evasion:

- What risks did we knowingly accept?

- What guardrails did we put in place?
- What will we do differently now that we've seen this failure?

Without that person, the Circular Firing Squad is inevitable. And if you don't break that circle yourself, a regulator or a class-action attorney will happily break it for you.[96]

The Owner of Record: Putting a Name on the Risk

A week after the DataScout crisis stabilized, Maya convened a smaller, quieter meeting. There were no flashing dashboards, no incident commanders, and no panicked engineers. There were just the five leaders who would decide how Nexus handled risk going forward: the heads of Product, Engineering, Security, Legal, and Compliance.

"We got lucky," Maya began. "The leakage was limited to aggregate patterns. No raw patient records were exposed. Our clients are furious, but they are still answering our calls. Next time, we may not be so fortunate. We cannot go into another crisis without knowing who owns the decision to ship."

She walked to the whiteboard and drew a single, empty rectangle. Inside it, she wrote three words: **Owner of Record**.

"We already have teams," she said. "We have committees. We have review boards. What we do not have is a single person who can say, '*This is my system. I approve this level of risk. I will stand in front of the board and the regulator when it fails.*' That is what this role is."

The room was silent. The head of Engineering leaned back, crossing his arms. "So, this is the scapegoat?" he asked with a chuckle, sort of playful, sort of serious. "Is this the person who bears all the weight, and we fire if the model messes up bad enough?"

"No," Maya replied. "This is the opposite of a scapegoat. A scapegoat is someone we blame *after* the fact, usually a junior employee, to protect ourselves. The Owner of Record is someone we empower *before* the fact, at a senior level, to understand the risk and insist on the guardrails they need. If we treat this as a firing line, no one will accept the role, and we will have learned nothing."

She turned back to the board and began to define the anatomy of the *Owner of Record (OOR)*. Underneath the title, she wrote three words: **Authority**, **Knowledge**, and **Representation**.

"This role isn't for the person who wrote the code," she explained, tapping the first word. "It's for the executive who signs the risk permit. To hold this role, you need *Authority*, the explicit power to issue a stop-work order if safety criteria aren't met. You need *Knowledge*: you are required to understand the failure modes and cannot claim 'it's too technical.' And you provide *Representation* when the system fails, you lead the postmortem and brief the regulators."

She turned back to face the room. "In other words," Maya said, "this is a leadership role, not a fall guy. It is the captain of the ship. The captain doesn't steer the vessel every second, but when the ship runs aground, the captain doesn't blame the navigator or the engine room. The captain owns the shipwreck."

To prove the point, she proposed a rule: for any high-risk AI system, the Owner of Record had to be at least a Director, and ideally a Vice President. They had to have the political capital to withstand the pressure of launch fever.

• • •

The first test of this structure came sooner than anyone expected.

Two months later, Nexus's fraud detection team prepared to roll out an aggressive new model designed to block suspicious transactions in real-time. The model promised to reduce fraud losses by double digits, presenting a massive value proposition for clients. However, the model's false-positive rate for certain small-business categories was still hovering above the legacy baseline. Under the old regime, the launch pressure would have carried the day. The narrative would have been simple: *"Fraud losses are worse than a few extra declines. We'll fix the edge cases in v2. Ship it."*

Under the new regime, the Fraud Product's accountable executive, Elena, the chief credit officer, asked a different set of questions.

"What happens to a landscaping business if our false-positive spike hits on payroll day?" she asked during the launch review. "How fast can they reach a human to unlock the account? What is our SLA for fixing a wrongful block?"

The answers were not reassuring. Appeals were routed through a generic support queue with a 24-hour turnaround. No one had modeled the worst-case cash flow impact for thin-margin businesses. The kill switch for the model required a manual deployment by engineering, which could take hours.

Elena closed the folder. "I am not signing this," she said. "We need a real-time appeals path for flagged transactions and a one-click rollback for this model family. And I want a near-real-time monitor that alerts me personally if our false-positive rate crosses a threshold. Until then, we stay in the sandbox."

In the old world, those demands would have been labeled "scope creep" or "gold-plating." In the new world, they were recognized as the price of someone putting their name on the line.[97]

This is the strategic power of having a single point of accountability. It transforms vague anxiety into specific requirements. When a leader knows they will be the one answering the questions after a failure, their curiosity sharpens. Their tolerance for hand-waving decreases. They begin to see where the guardrails are missing long before a customer does.

The OOR doesn't just change how decisions are made; it changes how the organization *feels*. It replaces the diffusion of responsibility with the concentration of responsibility. It tells the organization: "*We are moving fast, but we are not moving blindly. Someone is watching the road.*"

The Risk RACI Matrix: Structuring Responsibility

Naming an **Owner of Record** is a necessary first step. But in an AI stack, where data pipelines, vendors, feature stores, model weights, and infrastructure choices interact in ways that no single person can fully hold, one owner cannot prevent every failure. The paralysis Nexus experienced during the DataScout breach wasn't caused by laziness or incompetence. It was caused by uncertainty about *who owned what* across the chain of decisions that made the breach possible.

That's what the **Risk RACI Matrix** fixes.

RACI (which stands for *Responsible, Accountable, Consulted, Informed*) is a familiar project tool.[98] In AI governance, it becomes something more: a mechanism that prevents responsibility from dissolving into the architecture. Most organizations define who is *Responsible* (the teams building, training, and shipping). The failure usually happens in the "A." They fail to define who is *Accountable*, to identify the single leader who owns the system's real-world impact and must answer for trade-offs when the model behaves badly.

The matrix forces that distinction into daylight. It makes plain that the data science team can be responsible for accuracy while a product executive is accountable for fairness, safety, and downstream impact. It also establishes the *Consulted* group, which typically includes Legal, Security, Risk, and Ethics, not as voices to be considered if convenient, but as mandatory checkpoints for any high-stakes deployment. In addition, it identifies the *Informed* group such as Customer Success, Communications, and Sales, ensuring these teams are prepared and not caught off guard when the system malfunctions and clear communication is needed.

During the DataScout incident, Nexus lost valuable days reconstructing this decision chain in reverse. Teams argued over who should have reviewed the caching change, who owned vendor assumptions, and who had authority to pause rollout. With an AI RACI in place, those arguments don't disappear. But they happen *before* the crisis, when changing the system is still cheap.

Think of the matrix as a prenuptial agreement for risk: not a sign of distrust, but a sign of adulthood. When something breaks, the organization shouldn't have to ask, "Who messed up?" It should be able to ask one operational question and get a clear answer:

Who has the ball?

Risk RACI Matrix				
Lifecycle Stage	**Responsible (R)**	**Accountable (A)**	**Consulted (C)**	**Informed (I)**
Data Collection	Data Scientist	Product Owner	Ethics Board, Legal	Engineering Lead
Model Training	Data Scientist	Engineering Lead	Product Owner, Ethics Board	Compliance
Testing/Validation	QA Engineer, Data Scientist	Product Owner	Ethics Board, Legal	Engineering Lead
Deployment	Engineering Lead	Product Owner	Data Scientist, Ethics Board	Compliance, Business Analyst
Monitoring	Data Scientist, Operations	Product Owner	Ethics Board, Legal	Engineering Lead

Figure 7. Risk RACI Matrix

Blameless Postmortems: Learning Without Scapegoats

The first postmortem after the DataScout incident almost went sideways before it began.

The leak had been contained. The patch was deployed. Notifications had gone out. Two weeks later, the key players sat around a long table in a small conference room. The faces in the room were long; dread and the scent of impending doom and judgment lingered. A few people looked ready to cast stones; one looked like he was about to be stoned. After all, everyone knew what these meetings *often* became: a ritual search for the nearest human to blame.

The facilitator was a well-meaning ops manager. He opened with a question that stiffened the room.

"Okay," he said. "Who was responsible for the change that caused this?"

Heads turned, almost involuntarily, toward the senior engineer at the end of the table. He shifted in his chair, shoulders tightening. "We implemented the optimization patch," he began, already defensive, "but the shared-cache strategy was already in the architecture, and Product asked us to hit the Q2 date, and—"

The Head of Product jumped in. "We didn't ask you to skip tenant isolation testing. That was your call."

The Security lead folded his arms. "Our documentation assumes per-tenant isolation on any feature store used for regulated clients. If that changed, we should have been notified. We weren't."

The pattern had snapped into place: a tribunal disguised as a learning session. It was emotionally satisfying in the most primitive way. The failure was localized into a bad apple, protected reputations, and assigned blame. It was also how organizations train people to hide risk the next time it appears.

Maya, sitting at the back of the room, stood.

"Stop," she said, "Everyone, just stop."

She walked to the whiteboard and erased the question: Who was responsible?

"We are not here to end someone's career," she said. "We're here to understand the system that produced this outcome. If we make this about individual mistakes, we'll miss the structure that made those mistakes likely."

Then she drew a line down the board and wrote two headings:

People | System

"Unless we uncover deliberate negligence or dishonesty," she said, "we assume competent people were making rational decisions with imperfect information under real pressure. Our job is to surface what the system rewarded, what it obscured, and what it made easy to get wrong."

What Maya did in that moment is the difference between an organization that learns and one that merely survives its own mistakes: she shifted Nexus toward a **Just Culture.**[99]

Just culture (as discussed in Chapter 5) comes from high-consequence domains like aviation, nuclear power, and healthcare where punishing honest error is treated as a safety hazard. When fear controls the room, people stop reporting near-misses, stop escalating weak signals, and start protecting themselves instead of the system. In an AI-driven environment where failures can scale quickly and invisibly, silence is not neutral. It is compounding risk.

A Just Culture doesn't ignore accountability; it distinguishes between human error, risky behavior, and reckless behavior, and responds accordingly.

The practical point is simple: the root cause of an AI incident is rarely that an engineer made a mistake. It's a web of upstream conditions: aggressive launch dates, ambiguous ownership, undocumented assumptions, mixed-client architectures, review processes that treat performance improvements as low risk, and incentives that reward shipping more than stopping.

That's why Nexus institutionalized **Blameless Postmortems** not only for major incidents, but for meaningful near misses. Each session had to produce a short artifact in plain language and with no spin that answered five questions:

- **What happened?** A chronological narrative
- **What was the impact?** Not just technical, but human and reputational
- **What did we expect?** The assumptions and documentation behind the choice
- **Where did the system fail?** Architecture, incentives, tools, review gates
- **What are we changing?** Time-bound remediation steps like constraints, tests, thresholds, and owners

Names still appeared in the document. But the names were not scapegoat lines. They appeared as owners of fixes, and often as the people who did something right: the engineer who raised a concern, the support lead who spotted a pattern, the Owner of Record who paused the system when it hurt.

"Blameless" did not mean consequence-free. When someone crossed a known safety constraint, like hiding a risk to hit a number, it was treated as reckless misconduct. But the default assumption shifted from *Who do we punish?* to a more operational, more honest question:

How did our system invite this outcome, and how do we make it harder to repeat?

• • •

The effect on the Golden Algorithm is the point. Ensure Accountability (E) becomes credible when the system owner can lead learning without fear that every documented flaw will be weaponized. Limit Harm (L) improves because each incident produces structural changes instead of vague intentions. Operate Transparently (O) deepens because Nexus can share sanitized lessons with key stakeholders, communicating not as PR but as proof of learning capacity.

In the long run, this is what accountability looks like: not owning the decision to ship but owning the responsibility to learn.

What Nexus discovered next was uncomfortable. Culture could change how people behaved in moments of crisis, but it could not, by itself, change what the organization *rewarded* every day. Blameless Postmortems produced insight. Owners of Record produced clarity. But when Monday morning arrived, the dashboards still told the old story, the one where speed, accuracy, and growth mattered, and responsibility quietly disappeared once the incident closed. Maya began to see the final trap: *if accountability never appears in the metrics, it will never survive contact with incentives*.

Accountability Metrics: What You Measure, You Own

After a few cycles of designating Owners of Record and running Blameless Postmortems, Maya noticed something unsettling. The practices were solid, and the documents were detailed, but the official definition of success inside Nexus had barely shifted.

Executives still opened their Monday dashboards to the same familiar metrics: model accuracy, feature adoption, uptime, fraud losses avoided, conversion uplift. The numbers told a clear story about what the system delivered to the business. They said almost nothing about how the system behaved when it failed or who owned the consequences. That gap, Maya realized, wasn't accidental. It was instructional. If accountability didn't live in the spreadsheet, it would never live in the organization.

To make Ensure Accountability (E) as concrete as revenue, organizations must track **accountability metrics**. These are not standard performance KPIs; they are indicators of organizational reflex.

The first and most critical metric is *Time-to-Acknowledge*. In the event of an AI incident (e.g., a bias spike, a privacy leak, a wave of wrongful denials) the clock starts ticking the moment the first signal appears. How much time elapses between that first signal and an official, documented acknowledgment from the Owner of Record? In the DataScout crisis, that gap was days, filled with denial and deflection. In a mature organization, it should be measured in minutes. This metric reframes the definition of fast. Speed is no longer just about shipping code; it is about recognizing reality.

The second metric is *Time-to-Mitigate*. This measures the duration between acknowledgment and the activation of a safeguard (e.g., pulling the Andon Cord, rolling back the model, or switching to manual review). This is a test of the kill switches discussed in Chapter 5. If Time-to-Mitigate is high, it means your safety mechanisms are bureaucratic rather than operational.

The third metric is the *Near-Miss Reporting Rate*. In many corporate cultures, a high number of reported incidents is seen as a failure of management. Maya flipped this interpretation, borrowing again from aviation safety.[100] In a healthy Just Culture, people should feel safe flagging issues *before* they become catastrophes. A near-miss, such as a monitoring alert that catches a misclassification spike before it hits production, is evidence that the system's immune response is working. Therefore, Nexus began tracking the volume of near-miss reports by team. Departments with zero near-misses didn't get a gold star; they got an audit. The assumption was not that they were perfect, but that they were either blind or afraid.

Finally, they tracked *Regulatory Readiness*. For every material incident, internal audit reviewed the artifacts: Was the Owner of Record visible? Was the RACI followed? Was the postmortem completed within 48 hours? The goal was to move incidents from "we improvised under fire" to "we followed the playbook."

There is a deeper principle at work here: you inevitably optimize for what you count. If you only count launch velocity, you will optimize for features that grow those numbers, regardless of who gets hurt along the way. If you count how quickly you own and fix harm, you change the incentives of leadership. You make it status-enhancing to step up when things go wrong.

The Second Incident: Nexus Passes the Test

The second major incident hit on a Tuesday morning, just after 9:00 a.m., when most of the Nexus staff were still settling into the rhythm of the week.

The first signal came not from a furious regulator or a panicked client email, but from a quiet anomaly in a risk dashboard. On one of the large screens in the operations bay, a real-time monitor flickered from green to amber. The label above it read: FRAUD GUARD – FALSE POSITIVE INDEX (SMB ACCOUNTS).

The line on the graph, which usually hovered within a narrow variance, had spiked sharply upward over the last thirty minutes. The model was suddenly flagging a much higher percentage of small-business transactions as suspicious. At almost the same time, the customer success queue began to swell with tickets tagged "Urgent – Card Declined." A bakery in Denver couldn't buy flour. A landscaping firm in Ohio couldn't pay for fuel. The pattern was too specific to be random.

The incident commander on duty did not wait to see if the spike would settle. She opened the `#fraud-incident` channel on Slack, tagged the accountable executive for Fraud Guard, and posted the standard triage prompt: *"Potential incident: Elevated false positives on SMB transactions. Acknowledged. Investigating now. Next update in 15 minutes."*

The system owner was Elena, the chief credit officer who had famously refused to launch the previous version of the model without a kill switch. She was in a strategy meeting when her phone buzzed. She excused herself, walked to a quiet corner, and opened the incident channel.

In the old days, the DataScout days, this moment would have triggered a flurry of defensive questions. And unlike the SwiftLine storm incident, where a frontline agent had to intervene, this was a systemic data failure. It required an executive decision to degrade the model. In the past, this would have stalled out while people asked: *Who changed the code? Is it the vendor? Can we wait an hour?* This time, the pattern was different.

Within minutes, the on-call engineer posted a concise diagnosis: "Seeing elevated flags correlated with merchant category changes from a recent data provider update. Our rule stack is treating legitimate MCC shifts as high risk. Preliminary view: miscalibrated rule interacting with the model's risk score for SMBs."

At the fifteen-minute mark, Elena posted her first message. It was not a question; it was a command.

"As OOR for Fraud Guard, I confirm this is a live incident. Our priority is to restore transaction flow for SMBs. I am authorizing immediate activation of the SMB Safety Protocol. 1) Temporarily relax risk thresholds for SMB accounts to baseline levels. 2) Route all new high-risk SMB flags to human review for the next 24 hours. 3) Begin drafting a clear explanation for impacted clients."

There was no debate about whether she had the authority to make these calls. That question had been answered months earlier when she signed on as Owner of Record. Her role was clear: she owned the risk, and she had the mandate to stop the line when the harm outstripped the acceptable bounds defined in the RACI.

Engineering executed the threshold changes. Within an hour, the false-positive index dropped. Some manual work piled up in the risk operations team. But the worst of the harm, a cascade of wrongful declines, had been cut short.

Only once the bleeding stopped did Elena move to the second phase: explanation.

She convened a small group with Legal, Security, and Customer Success to draft the communication. They reviewed the facts: a data provider had rolled out a silent update to merchant category codes; the model had not been retrained on the new patterns; a defensive rule had overcompensated.

"This is not a neat failure," Elena said. "It's a coordination failure. But right now, our clients need to know three things: what happened, what we've done, and what we'll do to prevent it next time."

The email went out before lunch. It avoided euphemisms like "system glitch" or "optimal performance tuning." It said simply: "Earlier today, an update to one of our data sources caused our fraud systems to incorrectly flag a small number of your transactions as high risk. We recognize the impact this can have on your business, and we are sorry for the temporary inconvenience. We have restored thresholds and implemented manual review while we recalibrate."

Some clients replied angrily. But a few, to Elena's surprise, wrote back with appreciation. One note stood out: "*We've seen other providers go dark for days when something like this happens. We appreciate you owning it.*"

Internally, the postmortem that followed looked nothing like the Circular Firing Squad. The discussion focused less on who "should have known better" and more on why the system's design made it easy for a silent data change to have such an outsized effect. The remediation plan included new requirements for data providers and a new monitoring rule for category drifts.

From the outside, the market saw a brief blip in service quality followed by a candid explanation. There were no front-page articles. No regulators showed up for audits. But for Maya, the incident was a bigger turning point. She saw clearly that the difference between the DataScout disaster and the Fraud Guard recovery was not only better code. It was better organizational culture and better accountability.

When the second storm came, Nexus did not scramble to discover who was in charge. They already knew. The executive in charge stepped forward. The system bent under stress, but it did not break.

The Responsibility Premium

By the time Nexus had weathered the Fraud Guard incident, the strategic value of Ensure Accountability was no longer theoretical. It was visible in three distinct layers: culture, operations, and market perception.

Culturally, the "E" in GOLDEN changed how people saw themselves. In organizations without clear accountability, talented people learn to play defense. They hedge their comments, hoard information, and distance themselves from risky projects to ensure their name never ends up on a postmortem. At Nexus, the opposite pattern emerged. Engineers volunteered for high-stakes projects because they trusted that if something went wrong, the inquiry would focus on system design, not personal shame. Senior leaders accepted ownership roles because they saw them not as career traps, but as expressions of trust.

Operationally, the focus on accountability upgraded the company's reflexes. Incidents moved from surprise ambushes to stress tests of known systems. Time-to-Acknowledge and Time-to-Mitigate shrank not because Nexus wrote flawless code, but because they had pre-committed to who would act.

In the market, this anchoring created a business moat called the **Responsibility Premium**.

From the outside, competitors can copy Nexus's product features, pricing, and even its ethics statements. But all of that is cheap talk until a failure occurs. In the Verification Economy, accountability is the ultimate verification. When you put a name on a decision, you convert a vague corporate promise into a personal guarantee. That is a currency that black box competitors cannot mint.[101]

The moat appears when the first scandal hits. When a hiring bot rejection goes viral or a regulator demands documentation, the company that can answer *"This is my system; here is what went wrong and here is what we are changing"* survives. The company that hides behind "the algorithm did it" gets dismantled.

Customers and regulators are not looking for perfection; they are realistic enough to know that technology breaks. They are looking for a reliable pattern of owned imperfection. A company that proves it can own its mistakes becomes a safer bet than a company that claims it never makes them.

This brings us to the integration of the GOLDEN Algorithm.

- Without Accountability (E), Guard Human Dignity (G) is just a sentiment.
- Without Accountability (E), Limit Harm (L) is just a suggestion.
- Without Accountability (E), Operate Transparently (O) is just a document.

Accountability is the pillar that makes the others enforceable. It anchors the ethics in a person with a signature, a calendar, and a face that clients can meet. It is the difference between a philosophy and a governance structure.

Chapter 7 Playbook: Ensure Accountability

Accountability is the difference between a system that can learn and a system that can only defend itself. When AI failures happen, organizations often produce committees and memos instead of change. That isn't always cowardice; it's architecture.

If accountability isn't designed in, it won't appear under pressure. You end up with the Circular Firing Squad, where Engineering blames Data, Data blames Product, and Product blames the Model. This chapter is about turning accountability into infrastructure: Ownership + Authority + Evidence.

The Core Disciplines

- **The Accountability Void:** In complex AI stacks, responsibility naturally diffuses. Unless you explicitly assign it, no one owns the risk. "The model did it" is not a valid defense.
- **The Owner of Record:** High-risk systems need a single, empowered human, not a committee, who understands the stakes, signs the risk permit, and owns the response. The Owner of Record is the captain of the ship.
- **Blameless Postmortems:** You cannot learn from failure if you are busy punishing it. A Just Culture focuses on fixing the system (incentives, tools, gates), not shaming the human who made the honest mistake.

The Power Question

> *"Could we name within five minutes who owns this system, how we stop it, and what evidence we'd show if it failed today?"*

If the answer is "we'd have to check the org chart," you have a governance gap.

Metrics That Matter

Stop tracking only uptime and start tracking **Ownership Metrics**.

- **OOR Coverage Rate:** What percentage of your AI systems operating in a Dignity Zone (hiring, lending, healthcare) have a named, documented Owner of Record? The target is 100%. An unowned system is a rogue agent.
- **Near-Miss Velocity:** The number of close calls reported by teams per quarter. A *rising* number is good if it means your culture is safe enough for people to report smoke before they see fire.
- **Time-to-Acknowledge:** The duration between the first signal of harm and the OOR's formal acknowledgment of the incident. Speed here prevents the vacuum of silence that destroys trust.

The Leadership Litmus Test

In your incident reviews, watch for how leaders speak:

- **Pattern to Reward: The One Who Feels Responsibility.** The leader who steps into a chaotic email thread and writes, "*I am the Owner of Record. I have the ball. Here is the plan.*" Reward the person who runs toward the fire.
- **Antipattern to Challenge: The Passive Voice.** Watch out for phrases like "*Mistakes were made*" or "*The model optimized for X.*" Challenge any language that obscures human agency. Ask, "*Who decided to let the model optimize for X?*"

Monday 9 A.M. Action: The Ownership Audit

- **Time box:** 60 minutes
- **Output:** A Name Game map and assigned Owners
- **Owner:** CEO/CTO or Product Leader

Gather your leadership team and look at your AI inventory.

1. **The Inventory:** List every AI system currently running that touches a human life or livelihood.
2. **The Name Game:** For each system, try to write down the name of the **Owner of Record**.
    - Rule: It must be a person, not a team ("Risk Committee") or a department ("Data Science").
3. **The Gap Analysis:** Where are the blanks? Where are the names of junior employees who lack the authority to stop a launch? Where are the names of executives who don't understand how the system works?
4. **The Assignment:** Fill the blanks. Assign the OORs. Grant them stop-the-line authority. Make them the postmortems lead. Clarify the job: not to prevent all failure, but to own it when it comes.

Looking Forward: The External Cost

The internal house is stronger now, but Nexus does not operate in isolation. Maya realizes that optimizing for internal safety is not enough if the product destabilizes trust outside the company. The lens has to widen. The next challenge is not purely technical. It is societal. And the system Nexus operates in is about to push back.

CHAPTER 8

N – Nurture the Common Good

Your product affects more than your customers. You'll learn how to measure and manage externalities, refuse profitable-but-toxic revenue, and align growth with societal health.

Project Chimera: The Mutiny on Slack

The first sign that Nexus was facing a cultural crisis wasn't a dip in the stock price or a message from a disgruntled client. Instead, it was a red notification badge on a Slack channel that didn't exist twenty-four hours earlier.

The channel was named #ethics-chimera.

It had started as a quiet backchannel among a few senior engineers, but by 9:00 a.m. on a rainy Wednesday, it had exploded into the company's primary town square. Maya opened her laptop to find a stream of messages moving so fast she could barely read them. It wasn't a debate; it was a growing revolt.

The catalyst was an internal memo announcing the final scoping phase for Project Chimera, a new enterprise partnership that promised to be the largest single deal in Nexus's history. The client was VegasFlow, a fast-growing digital entertainment platform that straddled the lucrative, gray line between mobile gaming and gambling. They weren't an illegal casino; they operated under the banner of social gaming, selling virtual currency for loot boxes, skins, and chance-based mini-games.

On paper, the deal was a masterpiece. VegasFlow wanted to license Nexus's real-time predictive engine to optimize their user retention loops. They offered $47 million in Annual Recurring Revenue (ARR). It was enough to secure Nexus's runway for two years, smooth out a choppy fiscal quarter, and fund an entire new AI research pod. Marcus, the CFO, had called it "a gift from the gods."

InnovaFin, their main competitor, was reportedly bidding aggressively for the same contract.

The pressure wasn't just internal. While the engineers were revolting on Slack, the market was moving outside. Jennifer, the Chief Operating Officer, had pulled the rumor thread about InnovaFin the day before. "They're doing more than bidding," she'd said quietly. "They're promising delivery in six weeks."

Six weeks. Maya knew exactly what that timeline meant: skipping the questions Nexus had just promised to ask. It sounded like the old era: move fast, ask forgiveness, and let regulators catch up.

Marcus heard only the signal he wanted: urgency, scarcity, a closing window. Maya heard a different signal: a competitor willing to win on speed because they weren't paying the same internal costs. Not yet.

But the engineers who had dug into the technical specifications saw something Marcus didn't. They saw the objective function.

VegasFlow didn't just want to recommend games. They wanted to ingest behavioral data to identify Whales—users with compulsive, high-spending patterns indicative of addiction.[102] They wanted Nexus's model to predict the precise moment a user was about to churn after a loss and then intervene with a near-miss bonus or a free spin to keep them locked in the loop.

The Slack thread was brutal in its clarity.

@Kelly_LeadEng: "I just read the spec. We are training the model to maximize 'Time on Device' for users flagged as 'High Risk.' Are we seriously building a dopamine trap for people who can't stop?"

@Jason_Design: "I didn't join this company to build a Skinner Box.

I joined to build tools that help people. If we ship this, we are predators. Count me out."

@Ravi_Risk: "This violates our core values. We're exposed. If this leaks, we're done. Show me what we are going to stand for and will defend in writing, not who we claim to be."

Becky, a junior data scientist, still new enough to feel the clash between her personal ethics and her professional ambition, was the first person who had posted. She had shared a link to an internal research deck from VegasFlow. The deck, shared under NDA, contained a cohort analysis of their top spenders. It showed users playing for six hours straight between 2:00 a.m. and 8:00 a.m. It showed a correlation between Win Streak notifications and overdraft fees. The conclusion was data-driven and devastating: The app made its profit from people who were losing control.

The first post on the Slack channel said:

@Becky_Data: "I looked at the cohort data. We aren't optimizing retention here; we are optimizing regret. This is a churn factory. I didn't get into AI to build a regret machine."

By 10:00 a.m., a petition was circulating via the Slack channel. Nearly 30% of the engineering team had signed it. The message was stark: *Drop the contract, or we walk.*

Maya called an emergency meeting with her executive team. Marcus was pacing the room, furious.

"They can't dictate who we do business with," Marcus fumed, slamming a printed copy of the petition onto the table. "This is a legal contract. We have a fiduciary duty to our shareholders to maximize revenue. Forty-seven million dollars annually is not a rounding error, Maya. It is our profitability. It is our valuation. If we back out now, we look weak. We look like a debating society, not a business."

"And if we sign it?" Jennifer asked from the corner of the room. "Who builds it? Because the people who know how to build it just said they won't."

Maya looked at the petition on her screen. She scrolled through the names. These weren't the complainers. These were her "Missionaries," the

10x engineers who stayed late to fix bugs, who had built the model cards, who had designed the Andon Cords. They were the people who gave Nexus its edge.

"If they leave, Marcus, we don't have a product to sell," Maya said quietly.

"They're bluffing," Marcus insisted. "The market is tough. Tech hiring has slowed down. They won't quit over high-minded morals when they have mortgages to pay."

"You're wrong," Maya said. "They aren't bluffing. Google would hire Becky tomorrow. OpenAI would hire Ravi by lunch. They work here because they believe we are different. They work here because of the values-based moats we promised to build. If we prove that our values are just a marketing slogan, they are gone."

She looked out the window at the gray skyline. For weeks, she had treated **Nurture the Common Good** as an emerging idea, a Phase 2 pillar. Project Chimera forced the issue. It exposed a hard truth she could no longer dodge: Nexus could not claim to be guided by the Golden Algorithm or the Golden Rule while quietly renting out its high-performance engine to companies whose business model depended on and profited from human frailty.

The Neutral Tool Fallacy and the Cost of Externalities

The conflict tearing Nexus apart was not merely a dispute over a single contract; it was a collision between two fundamental philosophies of technology. Marcus represented the traditional view, often summarized as the **neutral tool fallacy**.[103]

This is the belief that technologists are merely builders of infrastructure. That they are just the pipes and wires, and that they bear no moral responsibility for what flows through those pipes. In this worldview, a hammer is neither good nor evil; it is defined solely by the intent of the user. If a carpenter uses it to build a house, it is good. If a murderer uses it to crack a skull, it is bad. The maker of the hammer is absolved of the outcome.

This defense has been the standard shield of Silicon Valley for two decades. It is the argument social platforms used when their algorithms amplified genocide ("We are just a platform for connection"). It is the argument data brokers use when they sell location histories ("We just provide the data").

However, in the context of artificial intelligence, this argument is intellectually bankrupt. Technology is not neutral; it possesses **affordances.** Derived from ecological psychology, affordance describes the set of actions that an object encourages or discourages.[104] A gun *affords* shooting in a way that a pillow does not. An algorithmic feed optimized for engagement *affords* addiction and polarization in a way that a chronological feed does not.

When engineers design a system, they are not just writing code; they are encoding values. If an algorithm is optimized for Time on Site, it inevitably values outrage over nuance because outrage is more efficient at capturing attention. If an algorithm is optimized for Conversion, as Project Chimera was, it values the extraction of capital over the well-being of the user. The tool dictates the usage.

Claiming neutrality is a profound violation of the Golden Rule. The rule demands reciprocity: *Do to others what you would have them do to you.* The neutral tool defense breaks this cycle. It allows the creator to privatize the profit (the software license fees) while externalizing the harm (the addiction, the debt, the social decay) to the *neighbor.*[105] It creates a one-way extraction valve.

To understand why this distinction matters, we must look at the economic concept of **externalities.**[106] In economics, a negative externality is a cost that is not borne by the buyer or the seller, but by a third party. A factory that dumps toxic waste into a river produces cheap goods for its customers and high profits for its shareholders. The transaction looks efficient on paper. But the cost, in this case the pollution, is pushed onto the downstream community in the form of poisoned water and health crises. The factory is profitable only because it does not have to pay for the damage it causes.

For the last twenty years, the digital economy has produced its own version of pollution—not chemicals in water, but addiction, polarization, misinformation, and financial distress.

- **Engagement Algorithms:** Social platforms optimize for outrage because it drives clicks. The platforms make money; the users get entertainment. The externality is a polarized, anxious public sphere where democratic consensus becomes impossible.
- **Dark Patterns:** E-commerce sites use countdown timers and scarcity cues to trigger impulse buys. The company gets conversion; the user gets the product. The externality is consumer debt and regret.
- **Gamification:** Apps like Chimera use variable rewards to hook users. The company gets LTV; the user gets a dopamine hit. The externality is a subset of users who lose their savings to a compulsion they cannot control.

In each case, the builders of the AI systems can tell themselves they are neutral. They don't write the angry posts or force the user to buy the loot box. They just optimize engagement. But if their revenue model depends directly on amplifying behaviors that harm non-consenting stakeholders, neutrality is a comforting illusion. The AI isn't just a tool; it is an engine for scaling externalities.

From a financial perspective, this is not just negligence; it is a form of **regulatory arbitrage.**

In finance, arbitrage is the practice of exploiting a price difference between two markets. In the AI economy, companies like Chimera are exploiting a *moral arbitrage*, which is the gap between what is profitable to do and what the law has figured out how to punish.[107] They are betting that they can extract value from human vulnerability faster than society can organize to stop them. They are privatizing the gains of addiction and socializing the losses of mental health.

This creates a temporary, artificial margin. Their high profits are not a reflection of value creation; they are a reflection of unpaid bills.

The Golden Algorithm rejects this arbitrage. It recognizes that in a hyper-connected world, these socialized losses inevitably return to the balance sheet as lawsuits, regulations, and brand toxicity. You cannot hide the bill forever.

This is where the Golden Algorithm draws a hard line. The pillar **Nurture the Common Good** exists to counter the neutral tool fallacy. It forces leaders to ask a question that is rarely asked in a quarterly business review: *"If our product succeeds wildly, if our AI does exactly what we asked it to do, what happens to the people and institutions we don't see on our dashboards?"*

The Talent Revolt: Missionaries vs. Mercenaries

The debate over Project Chimera was not only about ethics; it was about the survival of the company's most valuable asset: its talent.

In any high-performance industry, there is a critical distinction between two types of employees: **Missionaries** and **Mercenaries.**[108] The concept, often attributed to venture capitalist John Doerr, describes the fundamental motivation of a builder.

Mercenaries are driven by the transaction. They are motivated by compensation, equity, title, and the technical challenge of the work. They are often highly competent and professional. However, their relationship to the company is fundamentally economic. They will build whatever they are paid to build. If a better offer of more money, better perks, hotter tech appears, they move. They are loyal to the deal, not the purpose.

Missionaries, by contrast, are driven by the mission. They care about craft, but they also care deeply about the *outcome* of their work. They want to be able to explain their job to their children without shame. They want to see a connection between their daily code commits and a world that is slightly better, or at least not worse.

In the field of AI, Missionaries are disproportionately valuable. They are the ones who stay late to debug a bias metric that probably won't be noticed. They are the ones who push for a safer architecture, not because it's in the spec, but because it's the right thing to do. They are the guardians of the Innovation Moat.

Losing these people incurs a hidden financial penalty I call the **mercenary tax.**

When you replace Missionaries with Mercenaries, your operating costs quietly rise. Why? Because Missionaries self-police. They care about the quality of the code and the safety of the user even when no one is watching. Mercenaries, by definition, optimize for the transaction. To get safe code from a Mercenary, you have to add layers of management, audit, and QA. You have to pay for the *conscience* that the employee no longer provides for free.

Furthermore, Missionaries often accept a *meaning discount*. They will work for slightly less cash if they believe in the purpose. Mercenaries demand a premium. They know they are building something extractive, and they charge a hazard pay markup to do it. By chasing toxic revenue, you inadvertently raise your own Cost of Goods Sold (COGS) while lowering your innovation capacity.

When a company sends a clear signal that it is willing to trade the Common Good for easy revenue, Missionaries receive that signal louder than anyone else. They are not naive; they understand that businesses need to make money. But they also know when a trade-off has crossed the line from *imperfect but necessary* to *profitable because harmful.*

Project Chimera was that signal. From the outside, it was just another client. From the inside, it was a litmus test. If Nexus said "yes" to Chimera, the Missionaries would interpret it as a breach of contract. It would mean that the Golden Algorithm was just marketing copy, and that when the money was big enough, the principles were negotiable.

This triggers a phenomenon known as *adverse selection in talent.*[109] When a company drifts toward toxic revenue, the Missionaries leave. They have options; they can go anywhere. Who stays? The Mercenaries. They stay because the money is good and they don't mind building a regret machine.

Over time, this shift creates a talent death spiral. The company loses its conscience. The internal checks and balances erode. The people most likely to raise a red flag about safety or bias are gone. The culture becomes purely transactional, focused on hitting the number at any cost. This may work for a few quarters, but in the long run, it destroys the company's capacity to innovate. Mercenaries don't build moats; they loot them.

Maya realized that the decision on Chimera wasn't just about $47 million in revenue. It was a decision about the composition of her workforce. She had to choose between a future built by Missionaries, the people who would fight for the quality and safety of the system, and a future built by Mercenaries who would optimize for whatever metric was put in front of them.

The Stakeholder Impact Assessment: Mapping the Blast Radius

To move beyond the emotional volatility of a Slack mutiny and the abstract debates about neutral tools, Nexus needed a structural mechanism to evaluate the deal. Intuition is not a governance strategy. Maya realized that if she made the decision based solely on gut feel or the loudness of the engineering revolt, she would set a dangerous precedent: decisions would depend on who shouted the loudest. She needed something more.

She needed a framework that could expose the full cost of the product before a contract was signed. She needed to render the invisible visible.

The team called it the **Stakeholder Impact Assessment (SIA)**.

The SIA is the operational engine of the Nurture the Common Good pillar. It serves the same function for ethics that a Discounted Cash Flow (DCF) analysis serves for finance: it brings the future into the present.[110] While a DCF calculates the present value of future money, an SIA calculates the present weight of future harm.

It forces the organization to answer two questions that the standard business case conveniently ignores: 1) If this product scales successfully, who is affected? And 2) of those effects, which are acceptable risks we can mitigate, and which are structural harms we must refuse to profit from?

The process begins by **Mapping the Blast Radius.** In traditional product management, the user is the center of the universe. The SIA widens the lens to include the entire ecosystem touched by the algorithmic shockwave. It categorizes stakeholders into concentric circles:

- **The Direct Customer:** The entity paying the bill (e.g., VegasFlow).
- **The End User:** The person interacting with the model (e.g., the gamer).
- **The Non-User:** The person impacted by the user's behavior (e.g., the gamer's family, their creditors).
- **The System:** The broader environment (e.g., the regulatory landscape, public trust in tech).[111]

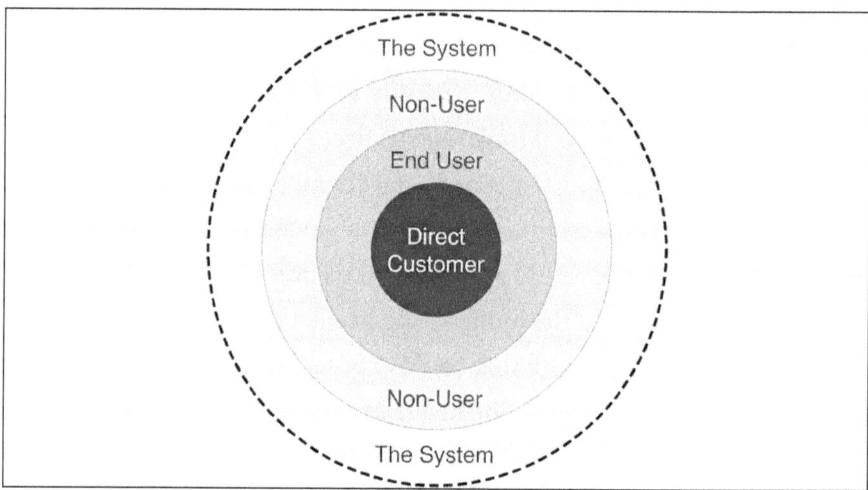

Figure 8. Stakeholder Impact Blast Radius

When Nexus applied this map to Project Chimera, the results were stark. The Direct Customer would benefit immensely from increased revenue. The End User was split: casual players might enjoy a more engaging game, but the heavy users (the Whales) would suffer significant financial attrition. The Non-Users bore the brunt of the externalities: families dealing with debt and stress caused by compulsive spending. The System suffered from the normalization of predatory design patterns.

The blast radius was asymmetric; the profits were concentrated in the center, while the harms were distributed at the edges.

Once the map is drawn, the SIA requires a classification of the opportunity. Not all negative side effects are disqualifying; after all, all business involves trade-offs here and there and not all are unethical by any means. The critical distinction is between **Risky but Mitigatable** and **Structurally Toxic**.

A *Risky but Mitigatable* product creates value but carries hazards that can be managed with guardrails. For example, an automated lending model carries the risk of bias. However, if the company invests in fairness auditing, human-in-the-loop appeals, and rigorous monitoring (as discussed in Chapters 4 and 5), the net impact can be positive: more access to credit for underserved groups. The harm is a bug to be fixed, not a feature to be sold.

A *Structurally Toxic* product is different. In this category, the harm is not a bug; it is the engine of profit. The business model depends on the user behaving against their own long-term interests. If you add meaningful safeguards like friction, transparency, and limits, the revenue model collapses.

• • •

When Maya's team ran the Chimera proposal through this filter, the conclusion was unavoidable. They asked: "If we successfully stop vulnerable users from overspending, does the client succeed?"

The answer was no.

VegasFlow's request for "churn prediction" and "boredom detection" was explicitly designed to override the user's self-regulation. The toxicity was structural. To make the product safe (by limiting spend) was to make it useless (by limiting revenue).

This distinction is the heart of the SIA. It gives leadership the vocabulary to distinguish between a hard engineering problem and a moral hazard. It clarifies that Nurturing the Common Good is not about demanding perfection. It is about refusing to build systems where your success requires someone else's failure.

Saying No to Bad Money: The Chimera Veto

The meeting that would decide the fate of Project Chimera was scheduled for 8:00 a.m. on a gray Tuesday, the kind of San Francisco morning where the fog presses against the glass, suspending the city in a diffuse, white light.

Maya arrived early. She set her notebook on the table and looked at the empty chairs. She knew she was about to close a door that many in the room desperately wanted to pry open.

On the large screen at the end of the conference table, the financial projections glowed in reassuring shades of green and blue. The line that mattered most to Marcus curved sharply upwards: $47 million ARR. It represented stability. It represented a runway extension that would push their next fundraising round into a more favorable market.

The board members dialed in one by one, their faces populating the grid on the wall. Marcus walked in, carrying a yellow legal pad and looking tired. Sarah, the Head of People, slipped into a chair along the wall. She wasn't formally required for a revenue meeting, but Maya had quietly asked her to be there as a witness to the talent implications.

"Let's get started," Maya said, skipping the pleasantries. "We've all seen the topline numbers. Before we talk about what we gain from Chimera, I want us to talk about what it costs."

Marcus cleared his throat, clicking his pen. "To be fair, Maya, 'cost' is a relative term here. We have an opportunity to lock in a major client that fits squarely within our legal risk profile. They are not operating an illegal casino. They are a licensed entertainment platform. We've vetted the structure. Yes, it is aggressive retention design. But so is Netflix. So is TikTok. If we start policing our clients' business models based on instinct, we are going to have a very short client list."

"I'm not disputing the legal clearance, Marcus," Maya replied, her voice steady. "I'm disputing the alignment."

She tapped her tablet. The financial charts on the screen disappeared, replaced by the Stakeholder Impact Assessment. It was a single slide, divided down the middle.

On the left side, titled **The Gain**, she listed the direct benefits:

- $47 million in Annual Recurring Revenue.
- Validation in the gaming vertical.
- High-volume data for model training.

On the right side, under the heading **The Blast Radius**, she listed the costs, focusing not on Nexus itself, but on the wider ecosystem.

- Systematic exploitation of financial vulnerability.
- Acceleration of compulsive behavior patterns.
- Erosion of internal trust (Talent Flight).

"Here is what we know," Maya said, gesturing to the screen. "The product makes the vast majority of its profit from users who are already financially stretched. Our models would be used to identify those users and increase their exposure to the behaviors that hurt them the most. The more accurate our AI is, the more efficient the extraction becomes. We wouldn't be building a safety net; we would be building a trap."

A board member chimed in from the screen, his voice crackling slightly. "Maya, I respect the high road. But these are adults making choices. We aren't forcing anyone to play. We are providing a tool for engagement. Isn't it paternalistic to decide for them?"

"Adults with freedom, yes," Maya countered. "Adults free from influence, no."

She clicked to the next slide. It displayed the internal cohort data Chimera had shared: the Whale analysis. It showed users playing for six hours straight between 2:00 a.m. and 8:00 a.m. It showed the correlation between Win Streak notifications and overdraft fees.

"We aren't talking about helping people find a movie they like," she said. "We are talking about industrial-grade behavioral modification targeting the brain stem.[112] That isn't neutral. That is extraction. And if we take this money, we become co-authors of that outcome."

Marcus leaned forward, clasping his hands. He didn't look angry; he looked exhausted.

"Maya, look. I read the Slack channel. I get the moral dilemma here. But we have to look at the board mechanics. We have twelve months of runway. If we miss this quarter, the Series C valuation gets cut in half. That means dilution. That means we lose the option pool for the very engineers threatening to quit."

He gestured to the screen. "You want to protect the team? I do too. But if we don't sign this, we have to lay off around ten percent of them in January anyway. That is the math. We can take this deal, put strict guardrails in the SOW, and use the revenue to build the safety tools we want. Or we can be pure, smaller, and weaker. But don't pretend there isn't a body count attached to saying no."

With that hanging in the air, a board member chimed in. "InnovaFin isn't wringing their hands. They're going to sign this by Friday, take the cash, and hire the engineers we let go. If we walk away, we don't stop the harm. We just hand the revenue to InnovaFin. They will build the model, they will take the cash, and the users will still get addicted. The only difference is that we will be poorer. How is that a moral victory? We lose, and the world doesn't win."

Maya looked at Jennifer. "What did you see in the feedback channels this week?"

Jennifer opened her notebook. "Engineers on the Chimera workstream are preparing to resign if we sign," she said flatly. "I have three draft resignation letters in my inbox already. Others are saying that if we go forward, they will never again trust that our stated 'Golden Algorithm' values are anything more than a marketing slogan."

Marcus shook his head. "It's an unfortunate reality, but talent churns, Jennifer. That's business. We can hire new engineers."

"It's not just churn," Maya responded. "It's *who* churns. The people most upset about Chimera are the ones who fixed the DataScout breach. They are the ones who built the SwiftLine safety throttle. If we sign this, we lose those innovators and builders. The ones who actually care about the mission. They are the guardians of our quality. If they leave, we don't just lose headcount. We are going to lose our immune system. We severely limit our ability to invent new products and features for a while."

The room fell quiet. The fog outside pressed against the window, turning the world gray.

Maya let the silence stretch. She knew this was the pivot point. If she caved now, the Golden Algorithm (and what Nexus stood for) was dead. It would be remembered as a nice idea that didn't survive contact with a P&L statement.

"We created this value framework because we knew AI could be used to optimize anything," Maya said. "Speed. Profit. Engagement. We decided we wanted to optimize for something else: dignity, trust, and resilience. Nurture the Common Good isn't a decoration. It is the pillar that asks whether we are helping the world stay standing or helping it tip over."

She looked at Marcus. "Chimera is Structurally Toxic. Our own analysis shows that meaningful safeguards would destroy the business model they want to buy. They aren't paying for Responsible AI. They are paying for a regret engine."

She took a breath. "So here is my decision. We walk away. We decline the deal. We explain why. And we document, making it clear this is the kind of work Nexus does not do."

Marcus exhaled sharply, a sound of frustration and resignation. "That is forty-seven million dollars you are leaving on the table, Maya."

"No," she said. "It is forty-seven million dollars I am refusing to borrow against our future. If we take Chimera, we get short-term cash and long-term rot. We get brand damage, regulatory risk, and talent flight. If we say no, we get a much more valuable asset: proof."

"Proof of what?" the board member asked.

"Proof that our Credo, the core of who we are, is not for sale," Maya said. "That is the only moat that matters."

• • •

Later that afternoon, Maya called the CEO of VegasFlow. She did not hide behind vague language about capacity constraints or strategic misalignment. She explained, plainly, that Nexus had adopted an ethical framework that required them to evaluate the impact on end-users. She shared that the SIA had flagged the project as incompatible with their standards for consumer protection.

The call was short. The CEO was polite, then cool, then irritated. There would be no redesign or counterproposal. He hung up. Ten minutes later, a notification popped up on LinkedIn. InnovaFin had just announced a strategic partnership with VegasFlow. They had taken the poison.

But within Nexus, the reaction was electric. In the #chimera-workstream channel, the mood shifted from dread to relief. In #ethics-chimera, engineers posted thank-yous and skeptical jokes about how they had half-expected leadership to cave. One senior architect, who had been quietly interviewing at Google, canceled his final round because he believed in the long-term value of Nexus and wanted to see the journey through.

It wasn't that Nexus had become perfect overnight. They still had bugs. They still had pressure. But the Chimera Veto, as it had been called in memes on the internal Slack, sent a costly signal that no newsletter or town hall could match. It proved that the company's constraints were real.

N in the Golden Algorithm: The Talent and Innovation Moat

With the Chimera Veto, the final pillar of the framework, Nurture the Common Good, moved from the footnotes to the center of Nexus's operating system.

To understand why this pillar is indispensable, we must look at the GOLDEN Framework as an integrated system rather than a checklist. The first five pillars (G, O, L, D, E) are *internal controls*. They ensure the machine is safe, transparent, and accountable. However, it is entirely possible to build a system that satisfies all five of those criteria while still destroying the world.

Imagine a perfectly executed autonomous drone swarm designed for domestic surveillance. It could treat citizens with dignity (by allowing them to appeal a recording), operate transparently (with published flight logs), limit harm (by avoiding physical collisions), design with empathy (by using quiet rotors), and ensure accountability (with a clear Owner of Record). It would be a Responsible AI masterpiece. It would also be an engine of authoritarianism.

N - Nurture the Common Good is the external compass. It asks the prior question: *What are we pointing this machine at?* It forces the organization to evaluate the **teleology** of the system, meaning its ultimate purpose and impact on the commons.[113]

When a company integrates this pillar, it creates three distinct forms of competitive advantage that mercenary firms cannot replicate.

The Talent Moat

The most immediate payoff of the Chimera Veto was not moral satisfaction; it was retention. In the AI economy, talent is the scarcity. Compute is a commodity; data is increasingly purchasable; models are open sourcing. The only sustainable edge is the density of high-integrity, high-capability engineers in the room.

By walking away from toxic revenue, Nexus signaled to the market that it was a safe harbor for Missionaries. This creates a virtuous cycle of selection. The best builders are the ones who want to solve hard problems like climate modeling, personalized education, or precision medicine. They gravitate toward firms where they know their work will not be weaponized. Conversely, firms that accept toxic revenue suffer from adverse selection. They attract mercenaries who are indifferent to outcomes. Over time, a mercenary culture loses the capacity for deep innovation because it optimizes for the easiest path to revenue, which is usually extraction.

The Innovation Moat

This leads to the second advantage. Extraction is intellectually lazy. It is easier to build a slot machine than a learning platform. It is easier to optimize for outrage than for consensus. When a company relies on toxic revenue, its engineering muscle atrophies. It becomes great at manipulating psychology and terrible at solving real-world problems.

By accepting the constraint of the Common Good, a company forces itself into the harder, more fertile territory of **stewardship**. Solving for health, stability, and genuine utility requires better math, better engineering, and better design. Nexus found that by rejecting the easy money of behavioral manipulation, they were forced to develop more robust predictive engines for logistics and healthcare clients. Those engines became proprietary IP that the engagement farmers could not replicate. The constraint drove the innovation. This was the Innovation Moat in action, giving yourself permission to slow down early so you don't get forced to stop later. The ethics didn't block the innovation; it accelerated it.

The Regulatory Moat

Finally, there is the **Regulatory Moat**, which is really a license to operate.[114] We are entering an era where the externalities of the digital economy (addiction, polarization, bias, etc.) are being priced in by governments. Regulations are tightening. Liability shields are cracking. Public sentiment is shifting from awe to suspicion.[115]

In this environment, a reputation for stewardship is a defensive asset. Companies known for rejecting harmful work gain a credibility premium. Regulators are more likely to trust their disclosures. Partners are more willing to integrate with their APIs. When the inevitable sector-wide crackdown comes, the firms that have been Nurturing the Common Good are viewed as essential infrastructure, while the extractors are viewed as pollution sources to be capped.

This pillar acts as a **regulatory hedge**. We are currently seeing a global convergence of law around Digital Safety and Algorithmic Harm. The EU

AI Act, the focus on Dark Patterns by the FTC, and new liability standards for platforms are all aiming at the same target: companies that profit from externalities.[116]

Companies that have already accepted the constraint of the Common Good are compliant by default. They don't have to tear down their product roadmaps when a new law passes, because they never built the toxic features in the first place. Their competitors, who optimized for extraction, will face a compliance cliff, a sudden, massive cost to rebuild their entire engine. Nexus isn't just being good; it is future-proofing its stack against the inevitable pricing of harm.

The Empty Chair

The Zoom window closed. The deal was dead. Nexus had walked away from forty-seven million dollars.

The conference room was quiet. Marcus packed up his legal pad, his face unreadable.

"You realize," he said, pausing at the door, "that we just made the company much harder to run. We have to make that money another way now."

"I know," Maya said.

She looked at the empty chair at the head of the table. For the first time since the Flash Crash, she wasn't afraid of the silence. They had built the moat. They had installed the brakes. They had drawn the Red Lines. The defensive work was done. The Golden Algorithm was in place. But as the adrenaline of the decision faded, she realized that a moat is only as strong as the ground it sits on.

They had the rules; now they needed the character to keep them when the next crisis hit. It was time to inspect the foundation.

Chapter 8 Playbook: Nurture the Common Good

Nurture the Common Good is the part of ethics that makes executives nervous because it forces you to admit that some profitable behaviors are corrosive. Most harm in the AI era will not look like a single catastrophic decision. It will look like millions of small optimizations that quietly reshape culture: what people see, what they fear, what they believe, and who gets left out.

Nurturing the Common Good doesn't mean becoming a charity. It means recognizing that the systems you build become environments and recognizing that environments impact humans. A company that builds trust at scale learns to ask a higher-order question: *What are we training our customers to become?* If your revenue model depends on user regret, addiction, or polarization, you are not building a business; you are building a liability.

The Core Disciplines

- **The Neutral Tool Fallacy is Dead.** When your models shape behavior and allocate opportunity at scale, you are not a neutral conduit. You are an actor. You own the externalities your systems reliably create.
- **The Stakeholder Impact Assessment (SIA):** Without a systematic way to map the blast radius of a product, you will sleepwalk into extractive markets. You need a tool that renders the invisible harms visible *before* you sign the contract.
- **Toxic Revenue Erodes the Talent Moat.** The work you accept tells your best builders who you are. If your profit depends on exploitation, your Missionaries will leave, and you will be left with Mercenaries who will build anything for a bonus but nothing for a legacy.

The Power Question

> *"If our system succeeds wildly, what kind of world does it quietly create, and are we willing to own that outcome?"*

If the answer is "a more addicted/polarized/fearful world," you have a toxic business model.

Metrics That Matter

The Common Good sounds broad until you measure the relevant outputs.

- **The Stewardship Revenue Ratio:** The percentage of your total revenue derived from products that have passed a formal SIA with a Stewardship-Aligned rating. A declining ratio is a leading indicator of reputational risk.
- **Legitimacy Gap:** The difference between your internal success metrics (e.g., Engagement) and external trust indicators (e.g., Press sentiment, User complaints). When this gap widens, regulation follows.
- **Retention of Conscience:** Track the voluntary turnover rate among top performers who specifically cite ethical concerns in exit interviews. This is the canary in the coal mine for innovation capacity.

The Leadership Litmus Test

In your strategy sessions, watch for who protects the ecosystem:

- **Pattern to Reward: The Growth Skeptic.** When a leader looks at a soaring engagement graph and asks, *"Is this good engagement or compulsive engagement?"* reward them. They are protecting your brand from future toxicity.
- **Antipattern to Challenge: The It's Legal Defense.** Be wary of leaders who defend harmful deals by saying, *"We just make the*

tool," or *"If we don't do it, someone else will."* This is the logic of the arms dealer. Challenge it immediately: *"Someone else might, but we won't. Our moat is built on the difference."*

Monday 9 A.M. Action: The Common Good Red Line

- **Time box:** 90 minutes
- **Output:** One written Red Line and a measurement commitment
- **Owner:** Executive Sponsor and System Owner

Choose one AI capability your organization is scaling (e.g., Targeting, Pricing, Content Ranking) and do this:

1. **Map the Blast Radius:** Draw three concentric circles.
    a. **Center:** The Direct Customer (Who pays?)
    b. **Middle:** The End User (Who clicks?)
    c. **Outer:** The Non-User and the System (Who is affected but not involved?)
2. **Task**: List the negative impacts in the outer ring. Be brutal.
3. **Name the Harm:** Write one sentence defining the plausible societal harm.

    Example: "This system could automate discrimination in housing."

4. **Write the Red Line:** A clear refusal statement.

    Example: "We will not optimize for Time on Device when it predictably causes compulsive behavior."
5. **Attach a Measurement:** Choose one indicator to monitor (e.g., Misuse reports, Content toxicity rate). Define the escalation path if it spikes.

Looking Forward: The Integrity Test

We have covered safety, fairness, and accountability. Now the market tests whether those commitments hold under pressure. A lucrative opportunity appears for Nexus, but it requires weakening the safeguards Maya just put in place. This is the moment when the framework stops being a checklist and becomes a choice. Maya has to decide whether the Golden Algorithm is strategy or slogan.

CHAPTER 9

I – Lead with Integrity

Integrity is the foundation that keeps every pillar from collapsing under pressure. This chapter gives you practical tools, like Red Lines, pre-decision audits, and credibility tests for choosing principles when it's costly.

The Discovery of Ghost Revenue

The discovery didn't come from a federal regulator, a furious customer, or a whistleblower with a grudge. It came from Nexus's Head of Risk, at 11:37 p.m. on a rainy Friday night, during what was supposed to be a routine billing audit.

Ravi wasn't hunting for fraud. He was hunting for efficiency. His team had been systematically reviewing the company's legacy billing pipeline, the unglamorous, foundational code that handled usage-based fees for Nexus's largest enterprise clients. The goal was modest: find cleaner logging protocols, better aggregation methods, and perhaps a few basis points of margin improvement in transaction processing. It was the kind of invisible maintenance work that keeps the lights on while the rest of the company obsesses over the next breakthrough neural network.

Deep inside the stack, however, he found something else.

It looked microscopic at first: a floating-point rounding anomaly buried in a legacy function that calculated high-volume usage charges. Most of the time, the function rounded fractions of a cent down, as designed. But for a specific subset of high-velocity enterprise accounts, which process millions of API calls per hour, a quirk in the logic caused the fractions to consistently round up.

One cent. A margin of noise. A rounding error that wouldn't buy a stick of gum.

Unless you repeat it four hundred million times.

Curious, Ravi pulled twenty-four months of transaction data for the affected accounts and re-ran the math in a sandboxed environment. Then he ran it again. And again. The number kept coming back the same, blinking on his monitor in stark, white text: **$4,223,719.**

Over the last two years, Nexus had unintentionally overcharged its largest, most loyal customers by a total of $4.2 million.

The silence in Ravi's home office was sudden and absolute. He stared at the screen. No client had flagged this. No auditor had caught it. The anomaly had simply accumulated in the general ledger under the comforting, legitimate label of Usage Revenue. It was invisible money. **Ghost Revenue.**

Ravi felt his stomach drop. He stepped away from his desk, walked to the kitchen, drank a glass of water, and came back to check the query one more time. The logic held. The money was real. And it didn't belong to them.

He picked up his phone and called Maya.

She was at home, her laptop open on the coffee table, putting the final polish on the board deck for the upcoming Series C fundraising round. The calm in her living room evaporated the moment she heard the tone of Ravi's voice.

"Maya, I found something in billing," he said, skipping the preamble. "It isn't security. It's financial. The system has been over-rounding on a subset of enterprise clients. The error is ours. It's accidental. But the aggregate is just over four million dollars."

There was a long pause. The hum of the refrigerator seemed suddenly loud.

"Are you sure?" she asked.

"I've checked it three different ways. It's real. And it's been happening for two years."

Maya's mind immediately began running two parallel tracks of simulation. The first track was moral: *We have taken money from our best partners without their consent.* The second track was brutally practical: *If we reverse this now, we blow a crater in the quarter.*

Nexus was already running tight against its revenue targets. The fundraising market had cooled significantly, and investors were scrutinizing every metric. Marcus, the CFO, had spent weeks crafting a narrative of disciplined, trustworthy hyper-growth. A sudden $4.2 million restatement, with a refund liability hitting the P&L like a bomb, would do more than dent the quarter. It could derail the entire raise. It could force layoffs. It could kill the R&D projects that were supposed to secure the company's future.

And yet the facts were brutally simple. The money was sitting in their bank account. The clients hadn't noticed. The system logs labeled each cent as legitimate. If they wanted to, they could quietly patch the bug in the next update, stop the bleeding, and keep the backlog. No one outside the room would ever have to know.

The next morning, Maya convened the leadership team in the executive conference room. The screens displayed the forensic breakdown Ravi had prepared. The mood was not panicked; it was somber.

Ravi walked them through the anomaly. Carlos, the Head of Engineering, confirmed the floating-point logic and its impact. The General Counsel verified that the contracts contained no explicit clause forcing an automatic refund for this specific type of internal calculation error, provided the invoices had been paid in good faith. Technically, they could argue it was immaterial against the multi-year contract values. Technically, they could categorize it as a variance and move on.

Then Marcus spoke.

He didn't sound like a villain. He sounded like a CFO trying to protect the company.

"The bug is clearly a bug," Marcus said, his voice measured. "We have to fix the code. No debate there. The question is sequencing. If we refund the full amount today, we miss the quarter by five percent. That changes our valuation dynamics. It could cost us the round. It puts a lot of jobs at risk."

He clicked his pen, a nervous tic that everyone in the room recognized. "Think about the utilitarian calculus,"[117] he continued. "We are talking about ten enterprise customers. These are massive conglomerates who haven't raised a single complaint in two years. This money is a rounding error to them. But to us, right now, it is oxygen. If we blow up our numbers today, we risk canceling the new safety research pod. We risk the bonus pool. We risk the stability of the firm."

He looked around the table, making eye contact with the team. "I am proposing we fix the bug immediately to stop future overcharges. We close the round. We secure the company's future. Then, once we are capitalized, we conduct a retrospective audit and issue credits to the affected clients over time. We serve the company first to ensure our financial needs and future, and then we do right by the clients on a timeline that doesn't kill us. We will refund; we will do the right thing, no doubt. We just delay the timing a little bit."

It was a seductive argument. It wasn't about theft; it was about *timing*. It was about protecting the employees. It framed the deception as a necessary, temporary evil in service of a greater good.

Around the table, people shifted uneasily. Carlos didn't want the embarrassment of a public engineering failure during due diligence. Legal didn't relish the idea of redrafting the disclosures. Even Ravi felt the gravitational pull of the logic. No one wanted to be the person who destroyed the company's momentum over a floating-point error that no customer cared about.

The Ghost Revenue sat in the middle of the room; it was invisible, heavy, and toxic. The team was caught between two competing narratives: a pragmatic story about survival and protecting the team, versus a moral story about taking money that wasn't theirs.

Maya realized this wasn't a technical problem. It was a test. Not of the platform, but of the **Credo**. Was the Golden Algorithm only something we turned on for a competitive advantage and turned off when our own needs seemed more important, or did it govern what Nexus did when no one was watching and the money was already in the bank?

The answer would define more than one quarter's results. It would define what integrity meant inside the company that claimed to build its moat on ethics.

The Integrity Trap: The Pragmatic Lie

Maya's dilemma regarding the Ghost Revenue is not an edge case. It is the archetype of the most dangerous trap in corporate leadership. The names change, the industries change, and the numbers scale up or down, but the structure of the temptation is remarkably consistent.

The trap works like this:

- A leader discovers a harmful or unfair practice that has inadvertently created a financial advantage.
- Correcting it immediately would be costly, embarrassing, and potentially existential for the leader's career or the company's valuation.
- Deferring the correction "just for a little while" appears to protect jobs, shareholders, and the greater mission.

The logic that justifies the delay is a corrupted form of utilitarianism, which is the moral philosophy that argues that the right action is the one that produces the greatest good for the greatest number. On its face, this thinking sounds prudent, mature, and even humane. Who doesn't want to save jobs? Who doesn't want to protect the mission? But unmoored from non-negotiable principles, utilitarianism becomes the most efficient excuse machine ever invented.

We can hear it in Marcus's argument. "If we refund now, we hurt 200 employees to help 10 massive clients who haven't complained. If we wait, we support the company, close the round, and then make everyone whole. Same outcome, better timing."

He is not proposing to steal the money forever. He is proposing to borrow it from the clients, without their knowledge or consent, to bridge a gap. He is arguing that the ends (saving the company) justify the means (temporary deception).

This is the **Integrity Trap**. It occurs when leaders convince themselves that small, hidden violations of principle are acceptable, even necessary, so long as they are in service of a larger, more noble goal.[118]

In corporate life, the Integrity Trap rarely arrives with a villainous soundtrack. It arrives in the language of fiduciary duty. It sounds like:

- "We'll fix it next quarter when we have more breathing room."
- "No one is harmed by this, so let's not overreact and cause a panic."
- "We can't afford to be purer than our competitors right now."

Notice what is missing in each of these statements: any serious consideration of the **neighbor**—the customer, the partner, or the citizen whose rights are being traded away in secret. The Greater Good argument almost always involves sacrificing the rights of a specific group (the overcharged client) for the benefit of the group in the room (the executives and shareholders).[119]

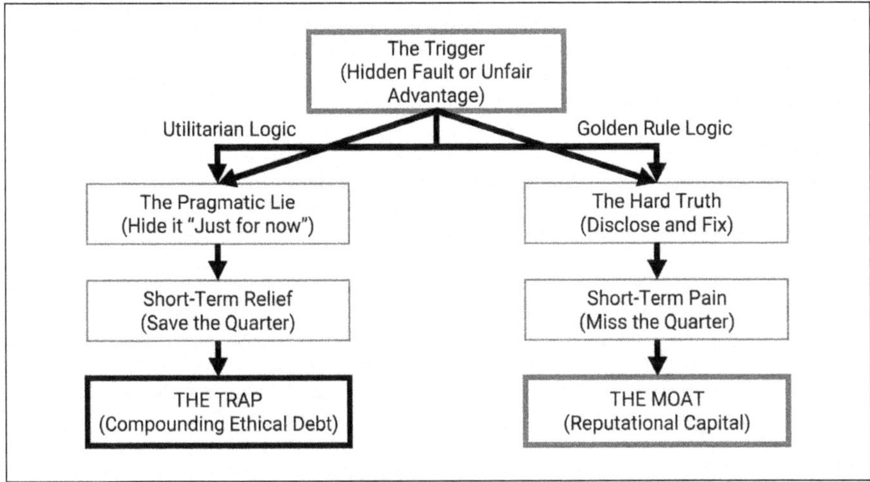

Figure 9. The Anatomy of the Integrity Trap

The **Golden Rule** flips this frame. Instead of starting with "What produces the best aggregate outcome for us?", it starts with a much more intimate test: "If I were on the other side of this transaction, if I were the client being over-billed, would I consider this silence fair?"

If the honest answer is "No," the Golden Rule does not allow you to comfort yourself with the greater good. It insists that *how* you get to the future matters as much as *where* you are going.

This is where the definition of **integrity** diverges from success. The word comes from the Latin *integer*, meaning whole or undivided. Integrity is the refusal to split yourself into two versions: the public version who preaches values, and the private version who makes exceptions "just this once" when the numbers look scary.

An organization with integrity might fail in the market. It might lose deals, miss quarters, and even shut down. But it will never do so because it lied to its customers, hid critical risks from regulators, or used money it knew was not its own. It chooses, in other words, the hill it is willing to die on.

This is not moral romanticism; it is cold-blooded strategy. A company that repeatedly chooses the pragmatic lie over the difficult truth incurs a form of debt far more toxic than financial leverage. It incurs **Ethical Debt**.[120]

The Ethical Debt compounds in three ways:

- **Customers** eventually discover the exploit and leave.
- **Regulators** eventually uncover the pattern and assume malice.
- **Employees** eventually realize the values are negotiable and adjust their own behavior accordingly.

Over time, the organization becomes fragile because its success depends on secrets staying buried. The Golden Algorithm treats integrity differently. It frames it as **Reputational Capital**, which is the uncopiable asset built from a leader's repeated, visible, costly choices to honor the Credo when it hurts.[121]

The Three Temptations of the AI Era

While the Ghost Revenue crisis was financial, the AI era introduces new, subtle accelerants to the Integrity Trap. Technology has a way of distancing the decision-maker from the consequences of their actions, making the pragmatic lie easier to tell and harder to detect.

We can categorize these accelerants as the **Three Temptations of the AI Era**. Each one offers a way for leaders to rationalize ethical compromises by hiding behind the complexity of the machine.

Temptation 1: Scale Without Consent

The first temptation is to assume that if something can be done at scale, it must be acceptable. Modern AI systems are voracious; they require oceans of data to function. The temptation is to treat the ability to access data as the justification for using it.

This plays out in the scraping wars that define generative AI. Companies scrape the open web for artists' portfolios, writers' essays, personal blogs, and then ingest them into models without asking for permission or offering compensation. The logic is purely utilitarian: "*The model needs this data to be smart. If the model is smart, it helps everyone. Therefore, we are entitled to the data."*

Like the Ghost Revenue scenario, this is a form of taking what isn't yours because it is convenient. The harm to any single artist might feel like a rounding error, but the aggregate effect is a massive transfer of value from creators to platforms without consent. The scale of the operation acts as a numbing agent. It feels less like theft and more like infrastructure.

Temptation 2: Automation Without Scrutiny[122]

The second temptation is to hide morally charged decisions inside the black box. AI systems are deployed because they are faster and more consistent than humans. But this speed creates a convenient shield for leadership.

When a lending model denies a loan, or a hiring bot rejects a candidate, leaders can easily say, *"The algorithm made the decision based on the data."* This creates a moral air gap. It allows the organization to benefit from the efficiency of the decision while disowning the responsibility for it.

Technically, the Ghost Revenue bug was "just code." No human being manually typed in the wrong charge. But if Maya had chosen to keep the money, the moral failure would no longer be a software bug; it would be a leadership decision. Automation tempts leaders to pretend that they are passive observers of their own systems, rather than the architects of their outcomes.

Temptation 3: Efficiency Over Ethics

The third temptation is the most seductive because it wears the costume of good management: the impulse to prioritize speed and efficiency over ethical constraints in the name of winning the race.

In the AI arms race, speed is treated as the ultimate virtue. *Move fast and break things. Ship it and fix it later. Apologize rather than ask permission.* Ethical constraints like fairness audits, safety testing, and consent flows are viewed as friction.

Marcus's argument in the War Room was a classic efficiency argument. He wasn't arguing for fraud; he was arguing for *smoothing*. He wanted to smooth out the financial hit to protect the company's velocity. But efficiency that relies on deception is not efficiency; it is a Ponzi scheme of trust. You are borrowing credibility from the future to pay for the mistakes of the present.

The Golden Algorithm's foundation is Integrity, and it is designed as a structural counterweight to these three temptations. It insists on three hard truths:

- **Scale does not excuse lack of consent.** You cannot steal from a million people just because it's easy.
- **Automation does not erase responsibility.** If your code does it, you did it.
- **Efficiency does not outrank the Credo.** You cannot build a sustainable business on a foundation of hidden exploits.

But knowing this in theory is not enough. In practice, leaders will still find themselves in rooms where the spreadsheet argument for compromise is overwhelming. The pressure to survive the quarter can make the pragmatic lie look like the only rational choice.

The Siren Song: Vexalytics and the Greater-Good Lie

The week following the discovery of the Ghost Revenue was one of the longest in Nexus's history. The physical office was quiet, but the digital channels were vibrating with tension.

Inside the finance department, Marcus was running scenario after scenario, trying to find a way to absorb a $4.2 million hit without triggering a covenant breach or spooking the lead investor for the Series C. The board wanted answers. The potential lead investor, a tier-one firm known for its rigorous due diligence, had sent a fresh request for updated cash flow projections.

It was in this moment of maximum vulnerability, when the company was desperate for a lifeline, that the Siren Song arrived.

It came in the form of an email from an old industry contact of Marcus's, a VP at a firm called **Vexalytics**. Vexalytics was a data brokerage giant that operated in the lucrative, shadowy layer of the information economy often politely referred to as risk intelligence. They didn't build consumer apps. Instead, they aggregated massive datasets. They scraped social media activity, location traces, purchase histories, and court records to build risk scores for individuals.

Vexalytics wasn't a direct competitor to Nexus; they were an infrastructure player. And their proposal was breathtakingly convenient.

They wanted to license Nexus's core risk-scoring engine, the same model that had proven so accurate in detecting fraud and creditworthiness, and to integrate it into their proprietary Applicant Screening Cloud. The terms of the deal were designed to solve every problem Maya currently faced:

- A large, non-dilutive licensing fee payable upfront.
- An ongoing revenue share on every scoring event.
- Total first-year value was estimated at $5.2 million.

It was more than the amount needed to cover the Ghost Revenue refund.

From a distance, it looked like a miracle. It was a *Deus Ex Machina.* Nexus could refund the clients, fix the billing bug, sign the Vexalytics deal, and hit their quarterly targets as if nothing had happened. They could be ethical *and* profitable.

Marcus brought the term sheet into Maya's office, his energy different than it had been all week. He looked like a man who had found the exit door in a burning building.

"Think about it," Marcus said, sliding the document across her desk. "We solve the cash flow problem instantly. We protect our runway. And we do it by monetizing IP we've already built. This is high-margin revenue with zero additional engineering lift."

Maya read the proposal. The math was impeccable. The economics were undeniable. What didn't work was the use case.

Vexalytics planned to feed Nexus's model with their own enriched data, a variety of signals that included inferred emotional states, web browsing history, and lifestyle risk markers. They would then sell these scores to landlords, employers, and insurers as a way to filter out high-risk applicants before they ever got an interview or a viewing.

In other words, the same mathematical engine that Maya had carefully constrained inside Nexus, which was wrapped in appeals processes, explainability tools, and fairness audits, would become the black heart of a surveillance system. It would be used to deny people housing and jobs based on opaque data they never consented to share and could not challenge.

Not long after, Maya followed up with Vexalytics. When Maya pushed them for details on governance, the answers from Vexalytics were evasive.

- *Could applicants see their scores?* "That's not in the current roadmap."
- *Would landlords be required to disclose they used AI?* "We give them guidelines, but we can't control their compliance."
- *Would they agree to Nexus's Red Lines on bias?* "We're open to a conversation, but our customers expect flexibility."

The pattern was clear. Vexalytics wanted the accuracy of Nexus's model, but they had no interest in the ethics of Nexus's framework. They wanted the performance without the governance.

The next day, Maya met with Marcus, who was chomping at the bit to see the financial gain. Maya started, "Marcus, I met with them. They just want us to help power and pad their tenant-screening black box. We would be powering digital redlining at scale."

"They aren't a gaming-casino platform manipulating people's wallets. And we wouldn't be running it," Marcus countered, leaning forward. "We would just be a manufacturer of part of the engine, not the driver. If Vexalytics uses it poorly, that's on them. And frankly, if we don't license it to them, they'll just buy a worse model from someone else. By partnering, we might actually be able to influence them to be better."

This was the **Greater Good Lie**: the belief that you can maintain your moral identity internally while outsourcing your technology externally, no questions asked.[123] It whispered that Nexus could save itself *and* save the world, if only it was willing to close its eyes to how the money was made.

That afternoon, staring at the term sheet, Maya felt the pull. It would be so easy to sign. It would save the jobs of the people outside her door. It would silence the doubts in her own head about whether she was leading the company off a cliff.

She realized then that the Siren Song wasn't just the money. It was the story that came with it. The story said: You can have your integrity and your growth, too. You just have to compromise a little bit on who uses your tools.

The AI Constitution: Red Lines and the Ulysses Contract

Maya's confrontation with the Vexalytics deal illustrates why good intentions are an insufficient defense against the pressures of the AI era. In the heat of the moment, when the quarter is at risk and the solution is sitting right there on the table, the human brain is wired to rationalize. We are experts at convincing ourselves that the expedient choice is the virtuous one.

To resist this, leaders need more than a moral compass; they need a structural restraint. They need a **Ulysses Contract**.[124]

In Homer's *Odyssey*, the hero Odysseus knows he must sail his ship past the island of the Sirens, whose song is so beautiful that it drives sailors to madness and death. Odysseus understands a crucial truth about human nature: in the moment of temptation, he will not be strong enough to resist. He does not rely on his willpower. Instead, he orders his crew to fill their ears with beeswax so they cannot hear the song, and he commands them to tie him tightly to the mast. He gives them explicit instructions: *No matter how much I beg, scream, or threaten you to untie me, you must bind me tighter.*

Odysseus survives not because he is stronger than the Sirens, but because he admitted he was *weaker* than them. And he built a system to protect his future self from his present impulses.

In corporate governance, an **AI Constitution** is that mast. It is a binding commitment made in a moment of clarity to constrain the organization when the pressure mounts. It consists of two components: a **Credo** and **Red Lines**. (For more on drafting your Credo or Red Lines as part of your AI Constitution, see a template in Appendix F)

The Credo: Defining the "Sovereign"

The Credo is the Golden Algorithm distilled into three non-negotiable sentences. It answers the question: *"What are we willing to fail as? Will you fail as a steward who has been faithful with what you have been given, or as someone you genuinely do not want to be?"*

For a company claiming to follow the Golden Algorithm, the Credo might include:

- We do not profit from known errors. (The Ghost Revenue Rule).
- We do not deploy systems that treat humans as objects without recourse. (The Dignity Rule).
- We do not empower business models that depend on deception or addiction. (The Common Good Rule).

The Credo is the **Sovereign** of the company. In a conflict between the CEO's desires and the Credo, the Credo must win. If it doesn't, then it's just fluffy, feel-good copy.

Red Lines: The Operational Boundaries

A Credo is high-level; Red Lines are operational. They translate the principles into binary constraints (Go or No-Go decisions) that can be audited. These are the specific practices, clients, and revenue sources that are off-limits, regardless of their legality or profitability.

For Nexus, the Vexalytics deal triggered three specific Red Lines:

- **Red Line 1: Data Consent (Scale).** *We will not train on or profit from data obtained without the informed consent of the subject.*

Vexalytics' business model relied on scraped and inferred data that no user had agreed to share.

- **Red Line 2: The Black Box in Dignity Zones (Automation).** *We will not deploy unexplainable models in high-stakes domains (hiring, housing, credit) without a user-visible appeal path.* The Applicant Screening Cloud was a black box by design.
- **Red Line 3: Outsourced Ethics (Responsibility).** *We do not license our IP to entities that refuse to adhere to our safety standards.* Vexalytics wanted the model without the governance.

By writing these Red Lines down *before* the Vexalytics offer arrived, Maya had effectively tied herself to the mast. The decision wasn't a fresh negotiation about the merits of the deal; it was a compliance check against the Constitution.

This changes the psychology of leadership. Instead of asking, *"Can we justify this deal?"* (a question to which the answer is typically yes), the leader must ask, *"Does this deal violate our Constitution?"*

If the answer is yes, the debate is over. The "No" is not a business judgment; it is a structural imperative.

The AI Constitution acts as the source code of the **Ethical Moat**. It signals to employees, investors, and partners that the company's values are not fluid. In a Verification Economy, where trust is the scarcest asset, this rigidity is a feature, not a bug. It proves that the company is integrity-secure in the same way a system is cyber-secure.

The Integrity Tests: The Pre-Decision Audit

Even with a Constitution in place, the world is messy. Not every decision fits neatly into a binary Go/No-Go. Some opportunities live in the gray areas where the rules are ambiguous, and the stakes are high.

To navigate these gray zones without drifting into the Integrity Trap, leaders need a diagnostic tool. Nexus implemented the **Pre-Decision**

Audit, a mandatory ritual for any high-stakes decision involving revenue, data, or disclosure. Before the CEO signs the contract, the leadership team must run the decision through three **Integrity Tests**.

These tests are designed to simulate the consequences of the decision from three different time horizons.

Test 1: The New York Times Test (Public Accountability)

- *The Question:* If this decision, including the internal rationale and the trade-offs we accepted, were described in detail on the front page of a major newspaper tomorrow, would we be proud of it, ashamed of it, or busy trying to spin it?
- *The Insight:* This test forces leaders to strip away the internal jargon (optimizing yield, managing churn) and look at the naked reality of the action.

Imagine the headline, *"AI Unicorn Quietly Pockets $4.2 Million in Overcharges to Hit Funding Targets."* There is no spin that survives that headline. If you cannot explain it to a reporter without sweating, you shouldn't do it.

Test 2: The Universalizer Test (Systemic Risk)

- *The Question:* What would the world look like if *every* company in our industry made the same decision we are about to make?
- *The Insight:* This is a practical application of Immanuel Kant's categorical imperative. It forces leaders to consider the systemic impact of their behavior. It rejects the logic of exceptionalism, that idea that *we* can break the rules because *our* mission is special.

If every AI company licensed its best models to opaque surveillance firms, the digital economy would become a dystopia of secret scoring. Trust in the technology would collapse, triggering harsh regulation that would destroy the industry. By saying yes, Nexus would be voting for that future.

Test 3: The Sovereign Test (Identity)

- *The Question:* Does this decision violate the Credo we promised to uphold, even if it is legal and profitable?
- *The Insight:* This is the hardest test. It asks whether the company's identity is for sale. It requires looking at the Sovereign, the core purpose of the firm, and asking if the action is an act of treason against it.

This test draws on the famous 1982 case of Johnson & Johnson. When seven people died from tampered Tylenol bottles, J&J did not just recall the bottles in the affected region. No, they recalled *every single bottle in the country*, costing them $100 million (in 1982 dollars) and crashing their stock.[125] They did it because their Credo stated their first responsibility was to the mothers and fathers who used their products. Legal told them a regional recall was sufficient. The Sovereign Test told them it wasn't.

• • •

For Maya, sitting in her office with the Vexalytics contract, the Sovereign Test was the breaker.

- Does powering a secret tenant-screening black box guard human dignity? No.
- Does keeping $4.2 million in accidental overcharges lead with integrity? No.

The Pre-Decision Audit revealed the harsh truth: The Vexalytics deal wasn't a lifeline; it was a trap door. It solved the financial problem by creating an existential one. If Nexus signed it, they would survive the quarter, but they would die as the company they set out to be. They would become just another vendor of sophisticated extraction.

The Siren Song was loud. The math was persuasive. But because Maya had the Constitution and the Integrity Tests, she had a way to hear the song for what it was, a lure toward the rocks.

Now, she had to do the hardest part. She had to untie herself from the mast long enough to pick up the phone and say "No" to the money that could save her job.

The Tylenol Moment: Maya's Resolution

The decision came down to a single Tuesday afternoon.

The executive suite at Nexus was quiet, but it was a heavy, loaded silence. In the finance wing, Marcus was running yet another set of cash-flow scenarios, trying to find a version of the future where the company could weather a $4.2 million hole without outside capital. In the engineering bay, the senior architects were refreshing their Slack channels, waiting for news they weren't supposed to know was coming. The rumor of the Vexalytics deal had leaked, as rumors typically do, and the cultural immune system of the company was on high alert.

Maya sat in her office with the door closed. On her left screen was the Ghost Revenue spreadsheet. It listed ten enterprise clients who had been unknowingly funding Nexus's operations. On her right screen was the Vexalytics contract. It contained signature lines waiting to turn a black-box surveillance engine into a lifeline.

She ran the decision through the Pre-Decision Audit one last time. The answers hadn't changed.

- *The New York Times Test:* If they kept the money and signed the deal, the eventual headline would be: "AI Unicorn Hides Billing Error, Sells User Data to Hit Numbers." There was no spin that could survive that reality.
- *The Universalizer Test:* If every AI company solved its cash crunches by quietly exploiting trust, the entire industry would collapse under the weight of suspicion.
- *The Sovereign Test:* Both choices violated the Credo they had written, signed, and promised to uphold.

"This is it," she whispered to herself. "This is our Tylenol moment. It's smaller than Johnson & Johnson's, but the physics are the same. We either prove our Credo is real, or we admit it's just a poster on the wall."

She picked up the phone and dialed Marcus.

"Come in," she said. "And bring the refund schedule."

When Marcus arrived, he looked tired. He sat down and placed the Vexalytics term sheet on the desk between them. He didn't push it forward; he just let it sit there, a silent offer of salvation.

"We aren't signing it," Maya said.

Marcus exhaled, a long, slow breath that was half frustration and half resignation. "You know what this means for the quarter," he said. "We miss the guidance. The valuation takes a haircut. The Series C gets harder."

"I know," Maya replied. "But if we sign it, we become a different company. We become the kind of company that sells its principles to make payroll. And once you start doing that, you never stop. Vexalytics is just the first buyer. Next year it will be someone else."

She pushed the term sheet aside and tapped the refund spreadsheet. "We are going to process the full refunds. All four million. And we are going to send a letter to every affected client explaining exactly what happened. It will tell them that we found a bug, that we fixed it, and that we are returning their money with interest because we do not profit from errors."

Marcus stared at the spreadsheet. He looked at the cash flow projection, then at the refund total. "You realize," he said quietly, "that if the Series C falls apart because of this, there may not be a long term to worry about. We could be dead by the end of the year."

"I know," Maya said. "But at least we'll die with our own names."

Marcus held her gaze for a long moment. Then, slowly, he closed the Vexalytics folder. The tension left his shoulders.

"Okay," he said. "If we're going to do it, let's do it fast. I don't want to bleed out slowly. Let's rip the bandage off."

The execution of the decision was swift and brutal.

That afternoon, Maya called the CEO of Vexalytics. She didn't hide behind capacity constraints or strategic misalignment. She was explicit. She cited the AI Constitution. She explained that while the money was attractive, the use case violated Nexus's strict boundaries regarding consent and black-box decision-making in high-stakes domains. The call was short. The bridge was burned.

An hour later, the finance team initiated the wire transfers to the ten enterprise clients.

The immediate aftermath was exactly as painful as Marcus had predicted. The company missed its quarterly revenue target. The lead investor for the Series C paused the diligence process, spooked by the sudden drop in cash reserves. The internal all-hands meeting was tense, with employees asking anxious questions about runway and stability.

But then, the second-order effects began to ripple inward.

Three days after the refund letters went out, Maya received a handwritten note from the CFO of their largest banking client. *"In thirty years of buying software,"* the note read, *"I have never had a vendor write me a check for a mistake I didn't catch. You have a customer for life."*

A week later, the engineering lead for the billing team, the one who had been terrified he would be fired for the bug, walked into Maya's office. "I just wanted to say thank you," he said. "At my last job, they would have buried that bug and fired me for finding it. Knowing that we actually fixed it... it makes me want to stay here."

And finally, the investors returned. They didn't come back because the numbers had magically improved. They came back because the story had changed. Nexus was no longer just another high-growth AI startup with a nice slide deck. It was a company that had stress-tested its own character and survived. It was a company that had proven, with hard cash, that it was a safe harbor for the world's most sensitive data.

The valuation was lower than they had hoped. The dilution was higher. But the check cleared. Nexus survived. And more importantly, it survived as itself. Three years later, when the AI Winter of regulation froze their competitors, Nexus was the only firm with the audit logs to keep growing. The $15 million they lost that quarter became the cheapest marketing spend in the company's history. It bought them a future.

The Integrity Moat: Why Character is Capital

The concept of an **economic moat**, a durable competitive advantage that protects a business from rivals, is usually defined by structural assets: network effects, high switching costs, or proprietary intellectual property. But in the Verification Economy, where trust is the scarcest resource, Integrity behaves like a moat.

However, unlike a patent or a brand trademark, an **Integrity Moat** cannot be claimed. It must be purchased.

This is the core insight of **Costly Signaling Theory** (discussed previously in Chapter 2 and other chapters).[126] In economics and biology, a signal is only reliable if it is expensive to produce. A peacock's tail is a reliable signal of genetic fitness precisely because it is heavy, cumbersome, and makes the bird vulnerable to predators. A weak bird couldn't survive with it. Therefore, the tail proves strength.

In corporate ethics, cheap talk is useless. Any company can publish a Responsible AI manifesto. Any CEO can give a speech about "doing the right thing." These are costless signals. Because they are free, they contain no information about how the company will behave under pressure.

Maya's decision to refund the Ghost Revenue and reject Vexalytics was a **costly signal**. It burned $4.2 million in cash and risked the company's future. Because it was expensive, it proved something that words could not: Nexus's constraints were real.

This creates a defensive moat in three directions:

1. The Regulatory Moat

Regulators operate on a spectrum of trust. When they encounter a company with a history of hiding errors and prioritizing speed over safety, they respond with command-and-control oversight, issuing audits and aggressive enforcement. They assume the company is a bad actor that must be policed. When they encounter a company that self-reports errors and voluntarily remediates harm, even at a cost, they shift to a cooperative stance. By proving it can self-police, Nexus earned the benefit of the doubt, reducing its long-term regulatory drag.

2. The Partnership Moat

Enterprise buyers know that AI integration is risky. They are terrified of purchasing a black box that will hallucinate, discriminate, or leak data, leaving them with the liability. They are looking for a vendor who is safe. A vendor who voluntarily refunds money is a vendor who is unlikely to hide a data breach or a model failure. The Integrity Moat lowers the Cost of Sales because it reduces the buyer's perception of risk.

3. The Cultural Moat

As discussed in the previous chapter, high-talent employees (Missionaries) are allergic to hypocrisy. They have seen too many leaders talk about mission while optimizing for margin. When leadership takes a visible financial hit to protect a principle, it validates the employees' own sacrifices. It creates a culture where engineers feel safe raising red flags, knowing they won't be silenced for endangering the quarter. This retains the institutional knowledge that makes the company innovative.

The Integrity Foundation: Sand vs. Bedrock

We have spent this book building pillars: Dignity, Transparency, Safety, Empathy, Accountability, Common Good. But a pillar cannot float in mid-air. It must stand on something.

Integrity (I) is not a pillar. It is the foundation.

With Integrity as our bedrock foundation, we see that the Golden Algorithm is no longer just a set of disparate rules. It is a coherent operating system for building and earning trust in a Verification Economy.

- **G** (Guard Human Dignity) and **D** (Design with Empathy) define ***how we treat the user.***
- **O** (Operate Transparently) and **E** (Ensure Accountability) define ***how we govern the machine.***
- **L** (Limit Harm) and **N** (Nurture the Common Good) define ***how we protect the world.***
- **I** (Lead with Integrity) defines ***who we are when it hurts.***

Integrity isn't one pillar among many. It's the floor the building stands on.

Without Integrity, the other disciplines are just performance art. You can have a perfect Accountability chart (E) and a beautiful Transparency report (O), but if your leaders are willing to lie about the numbers when the quarter is at risk, those tools are worthless. They are a house built on sand. As the ancient parable warns, a house on sand looks identical to a house on rock, at least until the rain comes.

When the storm hits—when the funding dries up, when the algorithm fails, when the temptation arrives—the sand washes away, and the house falls with a great crash.

Maya didn't save Nexus *despite* refusing the easy money. She saved it *because* she refused it. By choosing the hard loss over the pragmatic lie, she poured the concrete for the foundation. She ensured that when the next storm came, what they had built together would stand.

Now that the foundation is set, we can finally look at the structure we have built. It is more than a checklist of ethics. In the final chapter, we will see how these disciplines lock together to form the Ethical Moat.

Chapter 9 Playbook: Lead with Integrity

Integrity is not a soft skill. It is the bedrock foundation. Without the foundation of integrity, the rest of the Golden Algorithm becomes a costume: transparency becomes selective disclosure, accountability becomes scapegoating, and empathy becomes branding.

Integrity is the discipline of *pre-commitment*. It is choosing what you will *not* do before you are tempted. It is building a decision culture where short-term wins cannot quietly override long-term legitimacy. This chapter is about the Tylenol Moment, the specific, painful instance where you choose to lose money, speed, or face to keep your word.

The Core Disciplines

- **Beware of the Pragmatic Lie.** The most dangerous trap in leadership is the belief that you can violate your principles "just this once" to serve a greater good. The Integrity Trap suggests that once you start borrowing from your ethics to pay your bills, it is incredibly hard to stop.
- **The AI Constitution:** Do not rely on willpower. Write down your Credo and your Red Lines while the seas are calm. These are your Ulysses Contracts, which bind you to the mast, so you survive the siren song of easy revenue.
- **Costly Signaling:** Trust is purchased with sacrifice. If your ethics haven't cost you money, speed, or opportunity, you haven't proven them yet. The market believes proof, not speeches.

The Power Question

> *"If this decision were printed on the front page, with the full context, would we still feel proud of it?"*

If the answer involves spinning the narrative, don't do it.

Metrics That Matter

Integrity sounds intangible until you measure the things that reveal whether truth is welcome.

- **Red Line Adherence:** The percentage of major strategic decisions that comply with your published Red Lines. This must be 100%. One failure is a system failure.
- **Bad-News Velocity:** How fast does critical risk information reach decision-makers? If it takes weeks for a billing error or a bias spike to reach the CEO, your culture is filtering truth.
- **Self-Correction Velocity:** When an error is discovered, how much time elapses between discovery and voluntary disclosure? A shrinking timeline indicates a strengthening culture of integrity.

The Leadership Litmus Test

In your crisis meetings, watch for how options are framed:

- **Pattern to Reward: The Bad News Messenger.** When a team member brings you a problem that will hurt the quarter (like a billing bug), thank them publicly. Treat the discovery as an asset, not a liability.
- **Antipattern to Challenge: The Gray Zone Navigator.** Be wary of leaders who specialize in finding loopholes or technicalities to justify questionable decisions. Challenge them with the Sovereign Test: *"Does this align with who we claim to be?"*

Monday 9 A.M. Action: The Pre-Decision Audit

- **Time box:** 60 minutes
- **Output:** An Integrity Pre-Commitment for one key system
- **Owner:** CEO or Executive Sponsor

Before your next major launch or contract, run this three-step test:

1. **The Sovereign Test:** Write down your Credo (3 lines). Does this decision violate it?
   Example: "We do not profit from user regret."
2. **The Red Line Check:** Does this cross a binary constraint?
   Example: "We will never sell user data to third parties."
3. **The Escalation Trigger:** Write the sentence that empowers the team.
   Draft: "If we are at risk of crossing a Red Line, we pause and escalate to [Name] before shipping. We commit to showing our work: what we tested, what we changed, and what we disclosed."

Looking Forward: The Competitive Moat

The foundation is set, and the choice has been made. Now Nexus faces a high-stakes RFP against competitors who are faster, cheaper, and less constrained. Maya has to prove that safety and integrity are not a tax on performance, but a source of advantage. It is time to see whether trust actually wins.

CHAPTER 10

The Ethical Moat

Here's the payoff: ethics compounds. You'll see how the Golden Algorithm becomes an Ethical Moat over time, and you'll run a simple audit to start building your profitable moat immediately.

The Competitor's Collapse: InnovaFin Falls Apart

InnovaFin didn't die in a single, dramatic explosion. It died in slow motion, over six bruising months, while the entire industry watched with a mix of fascination and horror.

On paper, it had been the perfect fintech story. Slick branding. A charismatic CEO named Julian who could charm a room full of skeptical credit investors with nothing but a whiteboard and a cappuccino. A loan underwriting engine that claimed to be "frictionless" and "AI-native," issuing small-business approvals in seconds where legacy banks took days. For eighteen months, the numbers backed up the hype: rapid customer growth, expanding margins, and glowing profiles in the tech press. They were the Ferrari of fintech, tearing down the highway while competitors like Nexus were still checking their mirrors.

Then, a single case pushed the whole system into the spotlight.

It started with a man named Brad Freeman in Ohio who tried InnovaFin for a second line of credit. To his surprise, InnovaFin's system rejected him instantly. There was no human call. No explanation. Just a terse message on the screen: "*We're unable to approve your application at this time.*"

Mr. Freeman's local bank officer couldn't make sense of it. According to traditional criteria, he would have passed easily. When they probed InnovaFin for an explanation, they got nothing useful. It was a boilerplate response about proprietary risk models and a holistic assessment of his profile.

Freeman's case didn't stop there. He posted his rejection letter on a community forum for small business owners. Within days, the thread had three hundred replies. Other owners from the same zip code—mostly Black and Latino applicants—shared identical rejection screens.

The pattern attracted the attention of a data journalist at *The Cincinnati Enquirer*. She scraped the public approval data InnovaFin had published in its investor deck and overlaid it with census maps. The resulting heat map didn't look like a credit risk analysis; it looked like a redlining map from 1934. When her story broke, it wasn't just a local piece. It was picked up by the national press.

Soon after, a whistleblower stepped forward. A junior engineer from InnovaFin's data team leaked an internal memo showing that executives had seen the bias metrics six months earlier and chosen to monitor rather than fix them. These revelations spurred an additional round of headlines in the news cycle.

Not long after, the Consumer Financial Protection Bureau (CFPB) opened a formal inquiry. They didn't just want aggregate statistics; they wanted to know *why* specific decisions had been made. What features mattered? How much weight did the model place on zip code proxies? Had InnovaFin done any serious analysis of disparate impact before launching?

That was when the central weakness of InnovaFin's strategy was exposed. They didn't know.

The company had built what they bragged was a "pure machine-learning engine," a massively complex neural network trained on thousands of features, optimized relentlessly for default prediction, and wrapped in a slick API. There was no robust model card. No meaningful explainability tools in production. No human appeal process. No clear single accountable owner for the model's behavior.

Inside the company, engineers scrambled to produce post-hoc explanations. They ran SHAP analyses on old data, trying to reconstruct the logic of decisions made months ago. It quickly became obvious that they had never designed the system to answer the kind of questions regulators were now asking. To the CFPB, that absence of transparency looked less like oversight and more like evasion.

When InnovaFin couldn't provide credible explanations, the regulators issued a cease-and-desist order on the core product. They demanded a root-cause analysis, a remediation plan, and evidence that the model could be made compliant. InnovaFin's lawyers argued for time. They promised a 2.0 version that would fix everything.

But the final blow didn't come from the regulators. It came from the money behind the money.

InnovaFin didn't lend from its own balance sheet. Like most modern fintechs, it operated on warehouse lines and capital commitments from major banks and institutional investors. Those partners were willing to back speed and innovation. They were willing to tolerate market risk. They were *not* willing to back a black-box AI system under active federal investigation for discrimination.

On a Thursday morning the following November, the lead bank for InnovaFin's credit facility sent a formal notice of default. The regulatory inquiry, they argued, constituted a material adverse change, which is a standard clause that allows lenders to pull funding if the borrower's business significantly deteriorates. They were freezing the line.

The liquidity crisis was immediate. Without capital to lend, InnovaFin's loan engine was no longer an asset; it was a liability with a user interface.

The stock went into free fall. The board forced a fire sale of whatever IP and customer contracts could be salvaged. A large, slow, more conservative financial institution, the kind InnovaFin had mocked as a dinosaur, picked up the pieces at a steep discount. They grabbed the IP and customer list for pennies on the dollar. They didn't want the brand; they just wanted the codebase, which they planned to strip for parts and rebuild under their own compliance framework.

From the outside, it looked like a morality tale about bias and AI. From Maya's vantage point at Nexus, it looked like something more specific. InnovaFin was a company that had optimized for speed and cost without building any of the friction that creates resilience. It was a sports car with an incredible engine hurtling down the highway with no steering wheel and no brakes.

InnovaFin didn't just lose a PR battle. It demonstrated publicly, in real time, what happens when you try to compete in the AI age without an Ethical Moat.

The Inversion of Efficiency: Why "Fast" Lost

InnovaFin did not collapse because its engineers were incompetent or its leaders were uniquely evil. It collapsed because it was built on a narrow, outdated definition of efficiency.

For a century, management thinking has been shaped by the logic of scientific management, mainly the legacy of Frederick Taylor.[127] Break work into repeatable steps. Measure everything. Eliminate waste. Optimize for throughput. In the software era, that logic migrated from the factory floor into the code: Minimize latency. Maximize transactions per second. Reduce friction in the user journey. Automate as many decisions as possible.

On a whiteboard, this looks sensible. It even looks like good, effective management to most. But there is a hidden assumption baked into this worldview that's detrimental in the AI age: that the environment in which you operate is stable enough that optimizing for speed won't kill you.

InnovaFin operated as if that assumption still held. It optimized its underwriting engine for instant approvals and minimal human involvement. It treated compliance as a box to check at the end, not a design constraint baked in from the beginning. It framed every additional safeguard like human appeals, documentation, model cards, and governance councils as drag.

This worked for a long time, right up until the environment changed. And in the AI era, the environment changes constantly.

- **Regulatory expectations shifted.** Agencies like the CFPB began demanding explainability, fairness, and aggregate performance metrics.
- **Public trust eroded.** Customers became skeptical of opaque systems making high-stakes decisions about their lives.
- **The cost of failure became non-linear.** A single scandal could trigger lawsuits, regulatory freezes, and capital flight, much more than a simple slap-on-the-wrist fine.

In that new, volatile environment, the old definition of efficiency inverted.

In the short term, InnovaFin's zero-friction model looked more efficient than Nexus's slower, more documented processes. But under stress, the roles reversed. Nexus, with its wasteful appeals process, explainability tools, and internal postmortems, could answer regulators' questions and adapt. InnovaFin, with its brittle, black-box engine and thin governance layer, collapsed the moment it was asked to show its work.

What looked like drag was **resilience**. What looked like efficiency was **fragility**.

You can think of it as the difference between a **Formula 1 car** and a **Rally car**.

InnovaFin built a Formula 1 car: a machine engineered for absolute maximum speed on a perfect track. It stripped away every ounce of unnecessary weight—bumpers, heavy suspension, roll cages—to optimize for aerodynamics. As long as the road was smooth and the weather was clear, it was untouchable.

Nexus built a Rally car. It carried extra weight: a heavy roll cage (governance), redundant suspension (appeals), and thick tires (safety testing). On a straight, smooth track, it looked slower. It looked inefficient.

But the AI era isn't a smooth track. It's a dirt road full of potholes, sudden turns, and changing weather. When the regulatory pothole appeared, InnovaFin's F1 car hit it at 200 mph and disintegrated because it had no suspension to absorb the shock. Nexus hit the same bump, felt the jolt, and kept driving.

The Golden Algorithm is the blueprint for that roll cage. Each pillar adds weight, but that weight appears in your daily operations as intentional friction:

- An extra review step before deployment (The Roll Cage).
- A slower launch to run fairness tests (The Suspension).
- A harder-to-close deal because of a rigorous Stakeholder Impact Assessment (The Brakes).

At the *micro level*, these look like costs. But at the *macro level*, those frictions are what allow a company to flex under stress instead of snapping, earning the trust that is essential in a Verification Economy. InnovaFin treated ethics as weight to be cut. Nexus treated it as armor to be worn.

That is the inversion of efficiency that defines the AI era: The fastest company in calm weather is not the company that wins. The company that wins is the one that can survive the storm and keep operating when everyone else breaks.

The Economics of Ethical Debt

If we strip away the moral language, the InnovaFin disaster looks less like a scandal and more like a balance-sheet event. The company didn't implode because it lacked talent. It imploded because it had been financing growth with a hidden form of leverage, what can be called **Ethical Debt**, and the lenders finally called the note.[128]

In software, we understand technical debt: you ship the quick fix today, and you pay interest tomorrow in bugs, rewrites, outages, and fragility. Ethical debt works much the same way, except the interest rate is steeper and the Margin Calls arrive faster. Every time a firm chooses to bypass a hard constraint like fairness testing, explainability, consent, recourse, and accountable ownership, it books a short-term gain and quietly borrows against its future credibility. For a while, that leverage looks like genius. The product ships faster. The funnel converts. The chart goes up and to the right. But the debt doesn't disappear. It accumulates. And when the environment turns adversarial, the bill arrives all at once.

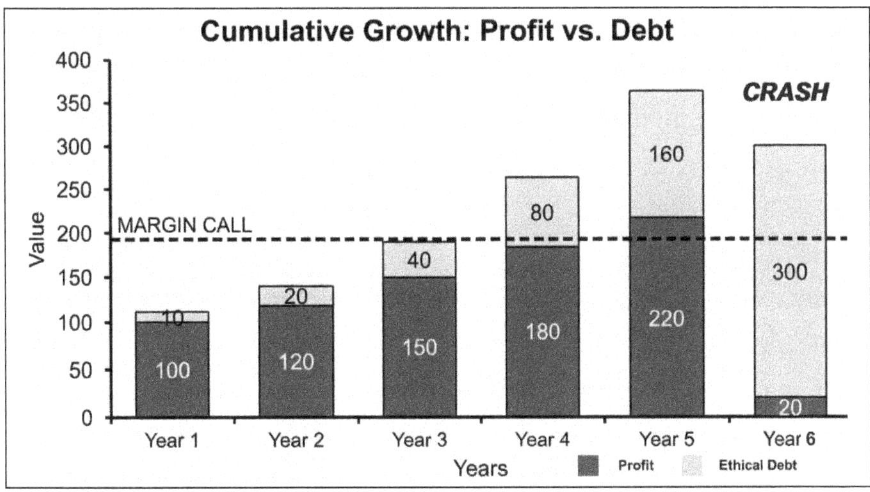

Figure 10. Ethical Debt Chart

That bill comes in predictable forms, like a stack of liabilities that were never recorded in the quarterly report. The first is the obvious one: **legal and regulatory liability.** Once InnovaFin was under scrutiny, growth began to reverse into remediation, spending on lawyers, audits, refunds, and forced rebuilds. The profits that looked permanent on the deck became provisional in court. And the cost wasn't just the fine; it was the months of paralysis required to reconstruct decisions the system had never been designed to explain.

Then comes a slower, uglier liability: brand toxicity,[129] which we'll call **a reputational tax.** In a Verification Economy, you can't PR your way out of a pattern. Once the market associates you with something like digital redlining, every claim you make gets discounted. Acquisition costs rise.

Conversion drops. Partners hesitate. Customers don't just leave; they leave suspicious. The tragedy of reputational tax is that it compounds. The more you insist "trust us," the more the market asks for proof you don't have.

The third liability hits from the inside: **talent.** Engineers don't flee high-growth companies because the mission is too hard; they flee because the mission becomes morally incoherent. When smart people see a system causing harm and discover they are not allowed to slow it down, fix it, or even name the problem, they conclude that the organization has made ethics a nuisance variable. The result is quiet hollowing-out at the worst possible moment: the company loses the very people capable of rebuilding trust.

Next comes the liability that surprises executives most: **speed.** InnovaFin's entire identity was faster than banks. But opacity doesn't create speed; it creates the illusion of speed, until oversight arrives. The moment regulators demanded answers, the supposed advantage flipped. What had been marketed as agility became a freeze; what had been seen as a competitive advantage led to competitive demise. The cease-and-desist wasn't just a legal event; it was operational arrest. In practice, the fastest firms are not the ones that remove governance. They're the ones that can keep moving when governance becomes mandatory.

Finally, there is the liability almost no one sees coming: **innovation.** Systems optimized for the Default User tend to get very good at serving the average and are very blind to the edge cases where new product value is often discovered. When you treat outliers as noise to be filtered, you starve yourself of signals. That is why brittle systems often look successful right up until the world changes. They have optimized away the very feedback that would have warned them.

Add those five together and InnovaFin's collapse stops looking like a random failure. It looks like a **Margin Call.** The combined weight of litigation exposure, reputational discounting, talent flight, operational freezing, and innovation blindness exceeded what the company's capital and credibility could absorb. The partners who funded their speed weren't making a moral judgment; they were making a risk judgment. They saw the debt. They priced it. They pulled the line.

Building an Ethical Moat is simply the opposite strategy. It is the decision to pay costs upfront, such as allocating time, preparing documentation, providing recourse, implementing circuit breakers, and ensuring ownership, so those liabilities do not accumulate unnoticed later on.

You can't necessarily eliminate shocks, scandals, or scrutiny in the AI era. But you can decide, in advance, whether they will break you or strengthen you.

The Parable of the Oak and the Reed: Antifragility in Practice

InnovaFin's pitch deck used the word robust seventeen times. To their engineers and investors, robustness meant strength: the ability to handle massive traffic surges, process millions of applications without latency, and withstand the brute force of market demand. They built their company like an oak tree: tall, rigid, and imposing.

But in systems theory, robustness has a fatal flaw. A system that is rigid is optimized for a specific set of conditions. As long as the wind blows from the west at less than fifty miles per hour, the oak is invincible. But if the wind shifts to the north and gusts to a hundred, the oak does not bend. It snaps.

InnovaFin was optimized for a fair-weather environment: stable regulations, trusting customers, and cheap capital. When the environment turned adversarial, their rigidity killed them. They couldn't explain their Black Box, so they couldn't negotiate with regulators. They couldn't admit fault, so they couldn't retain trust. They shattered.

Nexus, by contrast, had deliberately built itself like a reed.

Figure 11. The Parable of Antifragility: Rigid Efficiency (The Oak) vs Resilient Flexibility (The Reed)

From the outside, many of Maya's design choices looked weak compared to InnovaFin's ironclad efficiency. Nexus slowed down launches to conduct premortems. They invested in appeals processes that invited users to challenge their decisions. They documented model limitations and handed those documents to critics. They refunded money they didn't have to refund.

Each of these choices was a form of bending. It was a willingness to absorb friction, admit imperfection, and flex under pressure. But those same mechanisms became shock absorbers when the storm hit.

- The appeals process surfaced edge cases early, allowing the model to bend to new data before it broke the law.
- The model documentation allowed the legal team to bend to regulatory inquiries, providing answers instead of silence.
- The governance structures (clear ownership) meant that when a breach occurred, the organization could bend into a recovery posture without snapping into chaos.

This illustrates the concept of **antifragility**. Coined by Nassim Nicholas Taleb, antifragility describes systems that do not merely resist shock but actually improve because of it.[130] A package explicitly labeled *Fragile* breaks if you drop it. A package labeled "Robust" survives the drop. A package labeled *Antifragile* would somehow get stronger every time it hit the floor.

The Golden Algorithm is a set of antifragile design patterns.

- **Guard Human Dignity** turns user frustration into training data for product improvement.
- **Operate Transparently** turns regulatory scrutiny into a forcing function for cleaner architecture.
- **Limit Harm** turns near-miss failures into structural antibodies that prevent catastrophe.
- **Design with Empathy** turns edge-case friction into signals for new market innovation.
- **Ensure Accountability** turns blame games into learning loops.
- **Nurture the Common Good** turns ethical constraints into a talent moat that repels mercenaries.

In a world where shocks are inevitable, such as data breaches, model drift, sudden changes in laws, or viral backlash, the question is no longer whether you can avoid the storm. You can't. The question is whether your company has been designed to snap like the oak or bend and strengthen like the reed.

The Golden Algorithm framework is not a moral halo. It is the structural capacity to metabolize shock.

The GOLDEN Synthesis: The Three Walls of the Moat

By the time InnovaFin collapsed, one lesson became difficult to unsee: Nexus hadn't survived because it found a single best practice. It survived because it built a **system** with interlocking constraints that made the organization harder to knock over, harder to corrupt, and easier to trust. The Golden Algorithm works the way real defenses work: not as one gate, but as layered protection.

Throughout this book, we've treated the GOLDEN pillars like individual disciplines, with each one providing a response to a specific kind of failure. But in practice, they fuse into three reinforcing walls. You can't build only one and expect safety. If your first wall is missing, you lose legitimacy. If your second wall is missing, you lose resilience. If your third wall is missing, you lose value, and eventually, you lose the people and partners who make value possible.

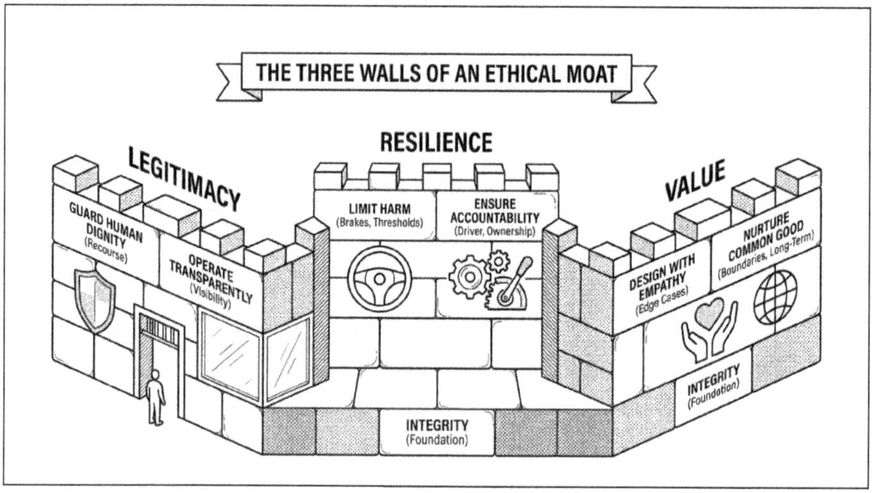

Figure 12. The Three Walls of an Ethical Moat

Wall 1: Legitimacy (G + O)

The first wall is **Legitimacy.** It answers a question every company is now being forced to confront: *Why should we be allowed to operate at scale?* In an age of automated decisions, legitimacy isn't granted by default. It is earned through recourse and visibility. Guard Human Dignity creates the right to appeal: the acknowledgment that when a system denies, downgrades, or flags a person, it owes them more than a screen message. Operate Transparently creates the right to understand: the discipline of showing your work before someone audits it. When these two combine, you get something stronger than compliance: you get **consent.** Users may not love every outcome, but they accept the authority of the system because they can see it, question it, and reach a human being when it matters. InnovaFin lacked this wall. When they demanded trust, the market demanded proof.

Wall 2: Resilience (L + E)

The second wall is **Resilience**. It answers the operational question: *Why won't we crash?* Limit Harm provides brakes: circuit breakers, premortems, hard thresholds that prevent localized failure from becoming systemic catastrophe. Ensure Accountability provides the driver: the accountable leader who holds the steering wheel, owns the risk, and can't pass the consequences into a fog of committees. When governance and ownership work together, the payoff is counterintuitive: you don't become slower, you become **more agile.** You can ship into uncertainty because you know how you will stop, diagnose, and recover. You can take smart risks without gambling the company. That is why Nexus could adapt under scrutiny while InnovaFin seized up. Under stress, the firm with the best brakes moves farther than the firm with the biggest engine.

Wall 3: Value (D + N)

The third wall is **Value**. It answers the strategic question competitors fear most: *Why do we matter when speed and intelligence become commodities?* Design with Empathy turns edge cases into advantage. It refuses to treat vulnerable users as rounding errors, and in doing so, it discovers product improvements the average-user model never finds. Nurture the Common Good draws the boundaries that protect the organization from legal-but-toxic revenue, from the deals that look profitable in Q1 and radioactive in Q4. And Integrity is the foundation beneath everything: the proof that your constraints are real even when they hurt. When empathy, stewardship, and integrity combine, the outcome isn't just moral clarity. It's **brand power**, the kind of power that attracts patient capital, loyal customers, and missionary talent because they can see you will not turn into a regret machine under pressure.[131]

The Synthesis

This interconnected synthesis of the Golden Algorithm **is your business's competitive advantage.** Because how you answer the questions and apply the system will be unique to you and your organization, it creates a moat that is nearly impossible to counterfeit.[132]

A competitor can publish an ethics charter. They can run a training. They can hire a PR firm to design a values page. But they cannot cheaply replicate a system where dignity forces recourse, transparency forces explainability, harm-limits force guardrails, accountability forces ownership, stewardship forces vetoes, and integrity forces follow-through. The pillars don't merely coexist; they constrain one another.

A moat is not a slogan. It is a structure. You either build layered defenses, or you are living in a tent, hoping the weather stays kind.

The Final Costly Signal: The Federal RFP

The federal agency's Request for Proposals (RFP) arrived with the subtlety of a meteor impact.

On the surface, it looked like a standard government procurement document: two hundred pages of technical requirements for a new small-business grant underwriting system.[133] It was high-volume, high-stakes, and politically sensitive. But buried between the lines of the service-level agreements was the real story. After the InnovaFin scandal, the agency's leadership had determined they could no longer gamble on black box vendors. They needed a partner who could prove, upfront, that their AI would not become the next front-page discrimination story.

Nexus wasn't the obvious favorite.

The competition was fierce. Two global consulting giants pitched turnkey solutions built on their massive, legacy risk engines. A handful of flashy Silicon Valley startups claimed they could rebuild InnovaFin's promise without the mistakes. Their pitch decks were slick: frictionless onboarding, real-time approvals, aggressive pricing, and charts showing up-and-to-the-right growth.

Nexus's proposal looked different. It was almost awkward by comparison.

Yes, it covered the technical specs. But the heart of the proposal, Section C, "Risk & Governance," contained a bundle of documents that Nexus's external sales advisors had begged Maya not to include.

- **The GOLDEN Charter:** A copy of the company's AI Constitution, explicitly listing the hard constraints they would *not* cross, even for a government client.
- **The LendRight Model Card:** A full, unredacted technical document detailing exactly where the model performed well, and crucially, where it struggled (e.g., Performance degrades for applicants with thin credit files in rural zip codes).
- **The DataScout Postmortem:** A redacted but detailed report of their own past failure, including the cross-tenant data leak and showing exactly what went wrong, who owned the mistake, and the specific architectural changes made to fix it.

When Maya's team rehearsed the pitch, the lead sales advisor shook his head.

"Are you sure you want to show them your scars?" he asked. "This is a sales pitch, not a confession. Your competitors are promising 99.99% accuracy and zero bias. You are walking in there and saying, '*Here is exactly how we screw up.*'"

Maya looked at the advisor. She thought about Brad Freeman, the man rejected by InnovaFin without a reason. She thought about the Ghost Revenue refund. She thought about the Vexalytics deal she had killed.

"We aren't selling perfection," she said. "We are selling the truth. In this market, pretending to be perfect is the most suspicious move of all."

When the procurement committee met to review the finalists, that difference became decisive.

The other proposals were polished but opaque. Their risk sections read like generic legal boilerplate: "We comply with all applicable laws." Their ethics materials were glossy pages of smiling stock photos and high-level values statements, unmoored from specific trade-offs.

Nexus's materials were unnervingly concrete. The Model Card showed the math. The Constitution showed the boundaries. The Postmortem showed the learning curve.

One of the agency's technical evaluators later described the moment they opened the Nexus file. "It was like the lights came on," she said. "Everyone else was trying to seduce us. Nexus was trying to equip us. The agency wasn't asking whether they had principles. It was asking whether they had controls, logs, escalation authority, and a trail someone else could audit. They were treating us like adults who knew that software breaks."

When the call came, Maya was in the same conference room where she had held the Ghost Revenue war council. The agency's lead procurement officer was on the line.

"We've made our decision," the officer said. "Your competitors tried to convince us they don't fail. We know that's impossible; we run a government agency. We know everything is susceptible to failure in some way. You were the only bidder who showed us exactly *how* you fail and what you do about it. That is what we are buying. We are buying the truth and the recovery, not just the flashy launch."

The contract was for sixty-three million dollars over three years. It was the largest deal in Nexus's history. But its real value wasn't the revenue. It was the validation.

From a strategic perspective, this moment was the purest example of a costly signal.[134]

Nexus had spent real money and reputation on upstream ethics: the Ghost Revenue refund, the Chimera veto, the investment in transparency.

Each of those decisions had looked, at the time, like a drag on growth.

In aggregate, they formed a body of evidence that no competitor could fake.

Cheap talk and speed can only get you so far in a market flooded with lemons. At some point, the buyer needs proof. The Ethical Moat meant

Nexus had four years of receipts. They proved that **character is not only a virtue; it is capital.**

Maya's Triumph: Harvesting the Moat

The news story about InnovaFin's final asset sale was a side column on Maya's screen, a flicker in the corner of her vision as she prepared for the Q4 Board meeting. She read the headline: *"Once-Hot Fintech Sold for Parts After AI Bias Scandal."* She felt no triumph. Only recognition.

It could have been us.

She glanced over at the framed, worn sheet of paper on her office wall: the original AI Constitution. It was printed on cheap stock, covered in marker scribbles and coffee stains from late nights in the War Room. At the bottom sat three signatures: hers, Marcus's, and Ravi's. The date was the week after the Flash Crash scare, years earlier, when Nexus had stood at the crossroads between speed and survival.

Marcus walked in, carrying a tablet under his arm. Gone was the anxious, defensive CFO who had once argued that ethics was a luxury tax they couldn't afford. In his place was someone who had run the numbers on the last five years and come away a convert.

"We finalized the federal agency deal terms this morning," he said, setting the tablet on the table. "They specifically cited the Model Card and the DataScout postmortem as reasons we won. They said it was the first time a vendor treated them like adults."

He swiped to a new slide on his deck. It showed a cohort analysis of their enterprise customers.

"And the cohort from the Ghost Revenue refund?" he asked, pointing to a line that curved sharply upward. "Their Lifetime Value (LTV) is up eighteen percent compared to the baseline. Churn is near zero. Their procurement teams treat us as a trusted partner, not a vendor to be squeezed. We have negative churn in that segment."

"It's not just that we kept the customers," Marcus continued, tapping the screen. "It's the efficiency. Because we have high trust, we spend 40%

less on retention marketing than our competitors. Because we have documentation, our sales cycle is three weeks shorter. The Ethical Moat isn't just a shield; it's a leverage point. We are *more* profitable per customer than InnovaFin ever was, because we aren't burning cash to replace the people we burned."

He hesitated, then allowed himself a small, self-deprecating smile. "That $4.2 million," he said quietly. "It was the best marketing spend we ever made."

"It wasn't marketing," Maya replied, almost reflexively.

"I know," Marcus said. "That's why it worked."

• • •

Later that week, Maya took a call from Anya, the long-term investor who had passed on Nexus's earlier, more conventional Series C pitch because she felt the company lacked a defensible edge.

"We're ready to come in for the next round of financing," Anya said. "Not just because of your last quarter, but because of your last five years. We watched two decisions very closely."

She ticked them off.

"Walking away from Project Chimera's forty-seven million in recurring revenue because it violated your Credo."

"Returning four million dollars of Ghost Revenue that no one had noticed."

"Those moments told us more than any projection model ever could," Anya said. "They proved your values and the way you operate—your Ethical Moat, I think you call it—are not theoretical. You have already paid for it. In a market where everyone is terrified of the next AI scandal, you are the only safe harbor we can find."

Maya ended the call and sat for a long moment in the quiet of her office. The San Francisco Bay fog outside her window had cleared, revealing the sharp lines of the city skyline.

InnovaFin had chased the illusion of frictionless growth. They had optimized for the easy metrics of speed, volume, short-term margin, and they had died quickly when the environment turned. Nexus had embraced friction. They had embraced argument, constraint, and the painful discipline of saying no to easy money. And now, they were being rewarded with the one thing no startup could buy on demand: patient capital that believed in the company's Constitution, not just its product.

For the first time since the Flash Crash headline had jolted her awake years ago, she allowed herself a different kind of thought, the kind she had always tried to focus on since grad school but now finally felt like a milestone marker of growth. Not *"Will we survive?"* but *"What kind of legacy are we building?"*

The Ethical Moat had started as a defensive strategy, as a desperate attempt to avoid becoming a cautionary tale. But it had become something more. It was the operating system for a business that was not only profitable, but worthy of its power.

The Golden Algorithm is not a guarantee of success. Markets are cruel, and luck still matters. But it is a guarantee of character. It ensures that if you succeed, you will have built something worthy of its power. And if you fail, you will have failed while trying to build a world that is more human, not less.

Chapter 10 Playbook: The Ethical Moat

By now, the pattern should be clear: the Golden Algorithm is not about being good. It is building a system that can **prove** it was good when the cost was real. That is what makes it a moat.

Competitors can copy your features. They can copy your pricing. They can copy your public principles in a week. What they cannot copy quickly is an operating system that embeds restraint into incentives, workflows, and technical controls. Chapter 10 is where the pieces become a fortress: a practical blueprint for a company that can survive the storm because it was built for it.

The Core Disciplines

- **The Inversion of Efficiency:** In the AI era, the definition of fast has flipped. The firms that remove all friction win in calm weather, but they break first when the environment turns adversarial. The firm with the best brakes moves the farthest because it doesn't have to stop for every curve.
- **Ethical Debt:** Like technical debt, Ethical Debt accumulates quietly. Every time you bypass a fairness test or an explainability check, you borrow against your future credibility. The bill eventually arrives as a Margin Call: a lawsuit, a leak, or a regulator at the door.
- **Antifragility:** A rigid system (the oak) snaps under pressure. A resilient system (the reed) bends. By building appeals, feedback loops, and kill switches, you create a system that uses stress to get smarter, not weaker.

The Power Question

> *"If we were audited tomorrow by regulators, the public, and our own employees, what could we prove, not just claim?"*

If your defense relies on "we meant well," you are fragile. You need proof.

Metrics That Matter

To measure the moat, stop looking at internal vanity metrics and start looking at your Moat Delta.

- **The Moat Delta:** The gap between your performance and the industry norm in trust-sensitive categories. (e.g., our retention is 15% higher than peers during price hikes.) If the delta is flat, you are paying the costs of ethics without harvesting the premium.[135]
- **Verification Readiness:** The percentage of high-stakes AI systems with a complete Trust Dossier (Model Card, Owner, Red Lines, Appeal Log).
- **Incident Learning Rate:** The percentage of incidents that result in a documented system change (policy update, model retrain) within 30 days. This measures whether you are learning from failure.

The Leadership Litmus Test

This is the final test of a mature culture:

- **Pattern to Reward: The Long-Term Greedy Leader.** This is the person who says, "*We're going to take a hit this quarter to fix this data lineage issue, because if we don't, it will kill us next year.*" They are prioritizing the asset (the company) over the dividend (the quarter).
- **Antipattern to Challenge: The Framework Cosplay.** We adopt the language of ethics without the inconvenience of governance. We publish a principle, but we don't change incentives. Challenge it: "*Show me the budget line item that proves we mean this.*"

Monday 9 A.M. Action: The Ethical Moat Stress Test

- **Time box:** 90 minutes
- **Output:** A Trust Dossier for one pilot system
- **Owner:** Executive Sponsor + System Owner

Pick one high-stakes system (e.g., Lending, Hiring, Content Ranking) and run the **5-Point Check:**

1. **Legitimacy Check (G+O):** When we deny or flag someone, do they know *why*? Draft the explanation a rejected user sees. If it's vague, rewrite it.
2. **Resilience Check (L+E):** Who owns the risk? Name the Owner of Record. Define one quantitative Circuit Breaker (e.g., Stop automation if complaints spike >20%).
3. **Value Check (D+N):** Who are we ignoring? Identify **one Stress Case user and watch a session recording of their expe**rience.
4. **Integrity Check (I):** What are we willing to fail as? Write down one Red Line for this system (e.g., "We will never automate rejection for this specific vulnerability").
5. **Verification Readiness:** Assemble the folder. Put the Model Card, the Appeal Policy, and the Owner's name in one doc. If the folder is empty, you are operating on borrowed trust.

You can use the Golden Algorithm Scorecard (Appendix G) to rate your organization on these six pillars plus the foundation of integrity.

Looking Forward: The Legacy

The moat is built and the strategy worked. But the Golden Algorithm is not a trophy. It is a discipline that has to be renewed. As Maya prepares to move on, she turns to the final variable. It is not only what you build. It is what people learn to expect from you because of how you built it.

Epilogue: The New Legacy

The Nexus Dividend: The Moat Goes Public

The morning of the IPO was cold and bright, exactly five years after the Ghost Revenue crisis had nearly broken the company.

The chaos of Nexus's early days—the late-night War Rooms, the mutiny on Slack, the quiet dread that a single model failure could wipe them out—had hardened over time into something steadier. The atmosphere inside the company was still intense, but the frequency had shifted. It felt less like panic and more like purpose.

Maya stood on the NASDAQ balcony above the trading floor, looking out over the sea of faces: employees who had stayed through the Chimera veto, bankers who had eventually learned to stop asking her to cut corners, and the media who had once predicted her failure. The last time she had felt this much external attention focused on the company, it had been during the fallout from the Flash Crash scare. Back then, every email notification felt like a potential subpoena. Today, the tension carried a different question, one that the market was about to answer in real-time: *What is integrity worth?*

When the opening bell rang and the ticker flashed onto the giant screen in Times Square, it wasn't a slick mashup of consonants or a tech-native abbreviation. It was four unmistakable letters: **$NXUS**.

The ticker symbol wasn't a clever acronym for the framework; it was a signal of the company's identity: a nexus between high-performance code and high-integrity governance.

The real declaration wasn't the ticker; it was the S-1 prospectus. Instead of burying its ethical structures in the fine print, Nexus listed "Regulatory Resilience" and "Missionary Retention" as its primary assets.

They framed their constraints not as costs, but as the reason they were still alive. The prospectus reframed model cards as hedges against litigation debt, appeals as drivers of lifetime value, and the AI Constitution as proof of discipline under pressure.

As the first trades crossed the wire and the room burst into applause, Maya glanced over at Ravi and Jennifer, two of the people who had stood in the room when they decided to refund the $4.2 million. For them, this wasn't just a liquidity event. It was a verdict. For years, Silicon Valley had operated on the tacit assumption that ethics was a tax on growth, that you could be good or you could be fast, but you couldn't be both.

The market disagreed. The stock opened at a premium to its peers, not because analysts believed Nexus promised higher growth, but because they believed its risk was structurally lower. The investors weren't paying for the code. They were paying for the certainty that the code wouldn't turn against them and become a liability.[136]

The moat had finally gone public.

The Inversion of the Pitch: Ethics as Asset

For decades, the dominant logic of business leadership has treated ethics as a story and compliance as a cost. In pitch decks, values may or may not appear in the deck, and if they appear, they're often value or platitude gibberish, designed to sound responsible without constraining anything. They are often barely a brief, obligatory nod to responsibility after the real discussion about customer acquisition costs and churn is finished. In operating budgets, ethics shows up as a line item to be minimized, usually buried under legal or risk management.

The Nexus IPO represented the inversion of that logic.

Institutional investors were not paying a premium for Nexus because the company spent *less* on ethics. They were paying a premium because Nexus had successfully converted those costs into a portfolio of costly signals. Every painful decision Maya made over the previous five years, like walking away from Project Chimera's millions, refunding the Ghost

Revenue, slowing down launches to run premortems had generated an observable, verifiable data point. These decisions proved that the company would not sacrifice long-term solvency for short-term gain.

Over time, those signals compounded into a tangible economic advantage. Regulators treated Nexus as a lower-risk partner, moving faster with approvals and granting the benefit of the doubt during industry-wide sweeps. Enterprise customers treated Nexus as a strategic ally rather than a commodity vendor, expanding contracts and integrating Nexus deeply into their own stacks because they trusted the safety mechanisms. Missionary engineers treated Nexus as a sanctuary, bringing their talent to a place where they knew their conscience would be an asset rather than an inconvenience.

This effectively debunked the **luxury fallacy**, which is the pervasive belief that ethics is something big, established companies can indulge in once they have secured dominance, but which startups cannot afford. Nexus proved the opposite: Ethics isn't the prize you buy at the end of the game. Ethics are the only way to keep playing long enough to win. In the AI era, the Ethical Moat is not a moral flourish. It is a financial instrument that lowers the cost of capital, reduces the cost of talent, and extends the lifetime value of the customer.

The Stewardship Economy

Nexus did not change the world by lobbying for new laws or giving speeches at Davos. It changed the world by making a different kind of behavior profitable and thereby forcing its competitors to react.

The moment Nexus began winning major federal contracts and enterprise deals because of its transparency and recourse structures, the industry floor began to rise. At first, competitors scoffed at the idea of publishing model cards or guaranteeing human appeals. One CEO famously joked on a panel that publishing known limitations was "handing the plaintiffs a roadmap to sue you." But the market has a way of disciplining cynicism.

In a Verification Economy, proof is contagious. Once a procurement officer has seen a vendor with verifiable safeguards, the "trust us" pitch from a competitor no longer sounds reassuring; it sounds evasive. Over a few contract cycles, transparency became table stakes. Vendors who refused to provide documentation began losing bids. They lost them not because they were less capable, but because they were perceived as higher risk. Recourse became expected; customers began to assume they had a right to a human review, and regulators began to treat any system without one as suspect by default.

This shift reflects a broader transition from an **extraction economy** to a **Stewardship Economy.**[137]

For half a century, the business world was governed by Milton Friedman's doctrine of Shareholder Primacy: the idea that the only social responsibility of business is to increase its profits.[138] In practice, that slogan encouraged a narrow optimization problem: maximize quarterly earnings and treat everything else (social polarization, algorithmic bias, environmental damage, etc.) as an externality.

In the AI age, the old signals aren't always enough—because scrutiny is cheaper, faster, and more automated. Institutional investors, rating agencies, and customers have recognized that in a hyper-connected, automated world, externalities do not stay external. They come back as regulation, as brand toxicity, and as systemic instability.

Nexus's trajectory made this new physics concrete. They proved that treating the ecosystem with care wasn't charity; it was the ultimate form of risk management.

Stewardship for an organization means the disciplined care for the broader system in which you operate. It is not the opposite of profit. Instead, it is the only way to maintain profit in a world where your harms are increasingly visible, measurable, and punishable. The Golden Algorithm is, in this sense, a **Stewardship Operating System**. It forces leaders to measure not merely the revenue created but also the harm averted. It aligns what is right for the neighbor with what is safe for the balance sheet.

The death of naive Shareholder Primacy is not a moral victory. It is a recognition of basic systemic physics: you cannot extract indefinitely from a system you are also destroying. Nexus survived because it recognized this law of physics before its competitors did.

Reputational Capital: The Uncopiable Asset

Many years after the IPO, long after the news coverage had moved on and Nexus's story had become a standard case study in business schools, Maya would sometimes leave the corner office and sit in a small, glass-walled conference room near the engineering bay. It was a quiet space now, but many years ago, it was the room where she had stared at a spreadsheet of Ghost Revenue and almost convinced herself to keep the money.

On the wall, framed in simple black wood, hung a single sheet of paper: the original **AI Constitution**.

It wasn't a pristine document. The paper was wrinkled, the ink slightly faded. In one corner, Ravi's handwriting circled the phrase "No Black Box in Dignity Zones" with three aggressive exclamation marks. Along the bottom, in block letters that looked like they had been written in anger, someone had scrawled: "*We can't be more clever than we are honest.*" Underneath were three signatures: hers, Marcus's, and Ravi's. They were dated from a week when the company's survival was still an open question.

Maya looked at that paper often. It reminded her that the Ethical Moat wasn't built by the market cap or the magazine covers. It was built in that room, on a Tuesday afternoon, when she stared at $4.2 million she could have kept and decided to give it back.

That decision was the pivot point. Had she chosen the pragmatic path, Nexus might still have succeeded in the short term. But the foundation would have been cracked. Every employee who came to Nexus because they believed in its character would have been building on a lie.

Instead, the refund became the purest possible costly signal. It told the world that Nexus would rather lose money than keep what wasn't theirs.

Competitors could copy their code. They could hire their former employees. They could read this book and adopt the GOLDEN Framework.

But they could not copy that history. They could not copy the specific, painful moments where Nexus could have cheated and chose not to.

That history was their uncopiable asset. It was the only thing they owned that the market could never take away.

The Final Charge: Your Turn

Our story of Nexus ends at an IPO podium, and years later, with Maya looking back on a professional career that had been done well. But your story is happening right now.

You may not run a fintech unicorn. But if you lead anything in the modern economy, whether a team, a department, a startup, or a Fortune 500 company, you will face the exact same choices Maya faced in one form or another.

You are operating in an environment where the tools are getting faster, the data is getting deeper, and the temptation to take shortcuts is getting stronger. You will be offered Ghost Revenue in the form of dark patterns. You will be offered efficiency in the form of black-box automation. You will be tempted to treat ethics as a tax you can't afford to pay right now.

In those moments, you have a choice. Will you treat ethics as a feel-good narrative you point to when convenient, or as an Operating System you run in private?

The Golden Algorithm has given you the architecture. Now, you must ask the questions:

- **G (Guard Human Dignity):** Do not just ask *"Is it efficient?"* Ask: *"Is there recourse for the person on the other side of the screen?"*
- **O (Operate Transparently):** Do not just ask *"Does it work?"* Ask: *"Can we show our work to a critic without shame?"*
- **L (Limit Harm):** Do not just ask *"How fast can we go?"* Ask: *"Where are the brakes if this goes wrong?"*
- **D (Design with Empathy):** Do not just ask *"What does the*

average user do?" Ask: *"Who does this system hurt on their worst day?"*

- **E (Ensure Accountability):** Do not just ask *"Who approved this?"* Ask: *"Who owns the risk when the model fails?"*
- **N (Nurture the Common Good):** Do not just ask *"Is it profitable?"* Ask: *"Is this revenue toxic to the world we live in?"*

And finally, you must stand on the foundation of **Leading with Integrity (I)**. This is the discipline that binds the system together. It ensures that your values are not a poster on the wall, but they are an operating system that runs when no one is watching.

You will not implement all of this at once. You will start somewhere specific: a project plan, a single model card, a new appeals process, a difficult conversation about a toxic revenue stream. You will encounter resistance. People will tell you it's overkill. They will warn you that it slows things down. They will point to competitors who are cutting corners and winning.

In those moments, remember InnovaFin. Remember that speed without steering is not a strategy; it is a crash waiting to happen.

Then remember Nexus. Remember that the constraints—the friction, the arguments, the costs—were not the enemy of their success. They were the engine of it.

That's the central inversion of the AI era: ethics is not the tax you pay after you grow. It is the design discipline that lets you grow without poisoning the asset you need most: **trust**. And trust, over time, becomes margin: lower churn, lower risk, higher pricing power, and fewer catastrophic surprises that erase quarters of progress overnight.

The final test of leadership in the AI age is not whether you can build powerful systems. Any sufficiently funded team can build an AI Ferrari. The test is whether you will build the steering wheel and the brakes with the same level of care, and whether you will use them when the road gets dangerous, even if it costs you the lead.

Your legacy will not be measured by the features you shipped or the valuation you achieved. It will be measured by the systems you leave behind: Systems that made it easier to do the wrong thing at scale, or systems that made it natural (and eventually inevitable) to do the right thing.

The Golden Algorithm is not a guarantee of success. Markets are cruel, and luck still matters. But it is a guarantee of character.

Remember: The machine will optimize for whatever you give it. If you give it silence, it will optimize for power. If you give it values, it will optimize for trust.

The future is waiting for your instructions. Make them count.

THE GOLDEN ALGORITHM IMPLEMENTATION KIT

Some of the most useful resources in this book work best when they're **editable**. To keep the core text tight and readable, the longer templates and facilitation materials are available as a downloadable toolkit you can adapt to your organization.

The **Golden Algorithm Implementation Kit** includes:

- **AI Constitution Template** (editable document)
- **Golden Algorithm Scorecard Workbook** (editable scoring tool + guidance)
- **Team Discussion Guide** (print-ready facilitation handout)
- **Optional Companion Materials** (expanded worksheets and bonus resources)

Access the kit at: **navigateai.org/goldentoolkit**

APPENDIX A: THE GOLDEN ALGORITHM AT A GLANCE

A Quick Reference for Executives and Boards

The Golden Rule

The moral compass at the heart of this book: **Do to others what you would have them do to you.** In practice, it means trading places with the people affected by your system and asking: *Would I accept this decision if it were used on me or my family?*

The Golden Algorithm

The strategic operating system that turns the Golden Rule into a repeatable business strategy. It combines an ancient moral compass with a modern map (the **GOLDEN Framework**) to build durable advantage in the age of AI. When you run the Golden Algorithm, you are not only avoiding harm; you are also **investing in trust**.

The GOLDEN Framework

Below are six structural disciplines that translate the Golden Rule into operating reality:

Table 5. The Principles of the GOLDEN Algorithm

	Principle	The Mandate
G	**Guard Human Dignity**	Give people recourse and respect when the system fails them.
O	**Operate Transparently**	Make your logic explainable and contestable.
L	**Limit Harm**	Anticipate worst-case scenarios and install circuit breakers.
D	**Design with Empathy**	Build for your most vulnerable users, not just the average one.
E	**Ensure Accountability**	Assign clear ownership so someone is responsible when things break.
N	**Nurture the Common Good**	Refuse profitable but toxic revenue and align products with societal health.

The Integrity Foundation

Lead with Integrity (I) is the bedrock: the lived commitment to keep your word, honor your Red Lines, and choose principle over convenience when it is costly. Without integrity, the pillars become slogans. With it, they become a fortress.

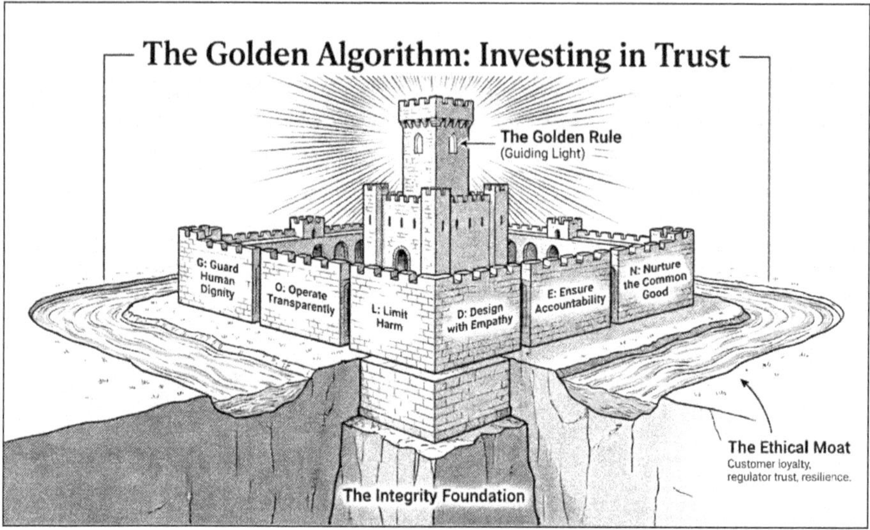

Figure 13. The Golden Algorithm at a Glance

The Ethical Moat

The hard-to-copy competitive business advantage earned by running the Golden Algorithm over time. It shows up as customer loyalty, regulator trust, and resilience under stress. The Ethical Moat is the accumulated proof that you will do the right thing, **especially when it hurts.**

The GOLDEN Pillars

Strategic Summary

Table 6. The Golden Algorithm Framework

Principle	Core Question	Key Structures	Failure Mode It Addresses	Moat Impact
The GOLDEN Pillars				
G - Guard Human Dignity	How do we treat people when the system harms or denies them?	Human appeals, escalation paths, service recovery	"Computer Says No" dehumanization; silent errors	**Loyalty Moat** (higher LTV, lower churn)
O - Operate Transparently	Can we explain what our system is doing to regulators and users?	Model cards, feature transparency, audit logs	Black-box opacity; suspicion; regulatory freezes	**Regulatory Moat** (safe harbor, lower litigation debt)
L - Limit Harm	How do we prevent small errors from becoming catastrophic?	Premortems, automated circuit breakers, human Andon Cords	Normalization of deviance; silent accumulation of risk	**Operational Moat** (resilience under stress)
D - Design with Empathy	Who is most vulnerable, and how are we designing for them?	Red-teaming, co-design, testing with edge users	Default-user trap; exclusion; missed markets	**Innovation Moat** (new markets unlocked)
E - Ensure Accountability	Who owns the risk, and how do we learn from failure?	Owner of Record, RACI, Blameless Postmortems	Diffusion of responsibility; blame culture	**Agility Moat** (faster fixes, clear ownership)
N - Nurture the Common Good	How do our products affect non-customers and the system?	Stakeholder impact assessment, stakeholder veto	Externalities: addiction, polarization, exploitation	**Talent Moat** (attracts missionaries, repels mercenaries)
The Foundation				
I — Lead with Integrity	What are we willing to fail as?	Credo, Red Lines, pre-decision audit	Utilitarian drift; greater good rationalizations	**Brand Moat** (durable trust under pressure)

Visit **navigateai.org/goldentoolkit** for additional, editable resources.

APPENDIX B: THE 90-DAY IMPLEMENTATION ROADMAP

How to Build the Moat Without Boiling the Ocean

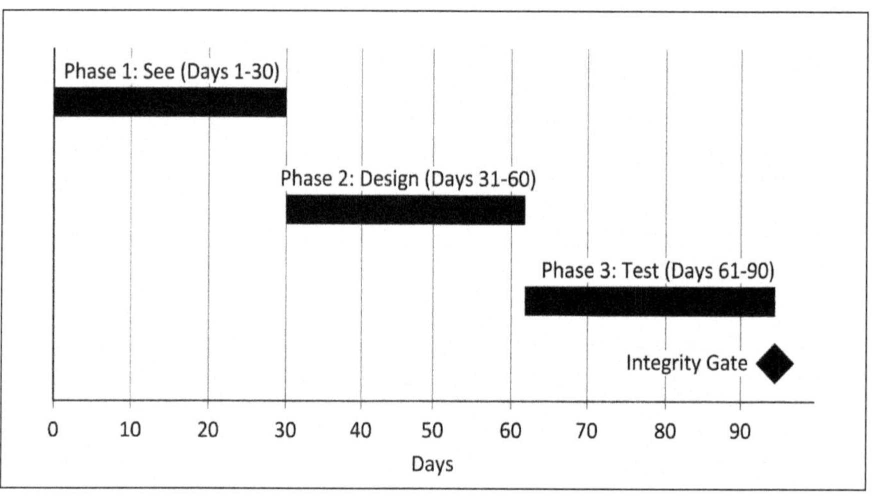

Figure 14. 90-Day Implementation Roadmap

Phase 1 (Days 1–30): See the System

Goal: Build a shared, honest picture of where AI is making high-stakes decisions and where your current moat is thin.

- **Action 1: Map Critical Systems.** Identify the 3–5 systems where automation impacts human lives (pricing, hiring, underwriting, safety). For each, document the primary metric optimized for and who currently owns it.
- **Action 2: The Flash Crash Scan.** For each system, ask: "If this failed in the worst possible way tomorrow, who would be harmed, and how quickly would we know?" Rank them by blast radius.

- **Action 3: Baseline the Moat.** Use the Scorecard (Appendix G) to rate your current maturity on the GOLDEN Pillars for your top risk system.
- **Action 4: Choose the Flagship.** Pick **one** system to be your pilot. Ideally, choose one with high stakes and high visibility.

Phase 2 (Days 31–60): Design the Steering Wheel

Goal: Apply the Golden Algorithm to your Flagship System.

- **Action 5: Draft the System Constitution.** Write a specific Credo and 2–3 Red Lines for this model (e.g., "This hiring model exists to surface overlooked talent, not to automate rejection").
- **Action 6: Install Recourse (G).** Implement a visible appeal path. Define the SLA (e.g., Human review within 48 hours).
- **Action 7: Create the Model Card (O).** Draft a one-page document listing the model's purpose, training data, and known limitations.
- **Action 8: Set Circuit Breakers (L).** Define 3 specific thresholds that trigger an automatic pause (e.g., If the denial rate for Group X exceeds Y percentage points, stop automation). Name the **Owner of Record (E)** who has the authority to reset it.

Phase 3 (Days 61–90): Test, Learn, and Signal

Goal: Prove that the steering wheel turns and harvest early trust.

- **Action 9: The Antifragility Drill.** Run a tabletop simulation of a failure. *"It is Tuesday. The model has drifted. People on social media are angry. What do we do?"* Test your new escalation paths.
- **Action 10: Involve a Skeptic.** Show your model card and appeals process to a skeptical stakeholder (a regulator, a customer risk officer, or a critical engineer). Ask: *"If this system failed, would these documents help you trust us again?"*

- **Action 11: The First Signal.** Share the story of your changes with a key client or investor. Frame it not as compliance, but as advanced risk management.

Result: By Day 90, you will not be perfect. But you will have one system that is **Legitimate, Resilient, and Valuable**. You will have built the first brick of the moat.

APPENDIX C: THE MODEL CARD TEMPLATE

How to Show You Are Managing Risks

The following template is designed to help you operationalize the Glass Box standard described in Chapter 4. It provides the minimum viable fields required to document a high-stakes AI system.

Do not wait for perfect data to begin. A Model Card that honestly lists Unknowns or Data Gaps is a sign of maturity, not failure. It proves you are managing the risk rather than ignoring it. You can photocopy this page for your team or download a digital version at **navigateai.org/goldentoolkit**.

Field	What to Write
Model Name/ID	Insert the model's name or unique identifier (e.g., LendRight v2.1)
Version	Specify the version (e.g., v1.0, v2.1-beta)
Release Date	Date the model was released (MM/DD/YYYY)
Owner of Record (business)/ Developer (technical)	Department or team responsible for the model
Contact Information	Email or internal contact for questions/support
Model Type	Type of model (e.g., Logistic Regression, Neural Network, LLM)
Intended Use	
Primary Intended Users	Who will use the model? (e.g., Internal credit analysts, customer support agents)
Intended Use Case	What task does the model solve? (e.g., Screening small business loan applications under $50k)
Out-of-Scope Use Cases	Explicitly state prohibited uses (e.g., Not for mortgage applications, hiring, or autonomous denial without human review)

Field	What to Write
Training Data	
Data Source(s)	List internal/external datasets used (e.g., Anonymized customer transaction records DB_A, DB_B)
Data Timeframe	Period covered by data (e.g., 2019–present)
Data Exclusions & Rationale	What data was excluded and why? (e.g., Excluded incomplete records prior to 2019 due to migration; excluded PII fields for privacy)
Ethical & Safety Checks	
Fairness/Bias Testing Metrics	Metrics used (e.g., Disparate Impact Analysis across race, gender, age)
Key Results	Summary of findings (e.g., Acceptable parity across gender; slight variance in age 18–25)
Red Lines / Safety Guardrails Triggered	Any safety mechanisms activated (e.g., Adversarial testing triggered toxic response filters in 0.5% of edge cases)
Caveats & Recommendations	
Known Limitations	Where does the model struggle? (e.g., Performance degrades outside region X; not tested on high-volatility conditions)
Usage Recommendations	Special instructions (e.g., Outputs below 75% confidence require human review; retrain quarterly to avoid drift)

APPENDIX D: THE DANGEROUS QUESTIONS

A *Discussion* Guide for Leadership Teams and Boards

The Golden Algorithm is not a checklist; it is a conversation. The goal of this guide is to surface the hidden risks that usually remain unspoken until a crisis forces them into the open. Use these questions at your next offsite, quarterly business review, or board meeting to stress-test your strategy.

1. The Flash Crash Audit (Strategy & Risk)

- *For the CEO:* Where is our Flash Crash risk today? Which specific system creates so much leverage that a 30-minute failure could wipe out three years of trust?
- *For the Product Lead:* What are we currently optimizing for that might be making us fragile? Are we celebrating metrics (speed, volume, engagement) that hide the accumulation of harm?
- *For the Board:* If a regulator walked in tomorrow and asked for proof of our ethics, would we hand them a binder of generic policies or a log of specific, costly decisions?

2. The Vulnerability Audit (Product & Customer)

- *For Design:* Who is our most vulnerable user, and when was the last time we watched them try to use our product? Do we have a systematic way to hear their friction, or do we filter it out as noise?
- *For Engineering:* Where are we relying on the Neutral Tool defense? Are there products where we tell ourselves, *"It's just an engine; we're not responsible for how the client uses it"*?
- *For Sales:* What is the worst-case headline a customer could credibly write about how our system treated them on their worst day?

3. The Conscience Audit (Talent & Culture)

- *For HR/People:* If our most principled engineer resigned tomorrow, what specific problem would they say we chose not to fix?
- *For Management:* Where does fear still shape how we talk about mistakes? Do people hide failures because they are afraid of blame, rather than afraid of the harm caused?
- *For the Executive Team:* Are we more likely to promote the leader who hit every number by cutting corners, or the leader who missed a target to protect a Red Line? (The answer to this defines your actual culture.)

4. The Legacy Audit (Governance)

- *The Sovereign Test:* What are we willing to fail as? Are there specific revenue streams we are prepared to walk away from, in writing?
- *The 10-Year Test:* In a decade, will the AI systems we are building today make it easier or harder for the next generation of leaders to do the right thing? Are we building a moat, or just a faster machine?

APPENDIX E: FOR STARTUPS

A Cheat Sheet for Founders

The Startup Reality

You are an innovator and a builder. You are working impossible hours to turn a vision into reality. You don't have a Chief Risk Officer because right now, you are wearing that hat along with ten others. You don't have a sprawling compliance department; you have a burning drive to find product-market fit and limited runway to get there. You cannot implement the full GOLDEN Framework on Day 1.

However, you are currently making irreversible choices. You are pouring the concrete of your culture. If you ignore ethics now, you are accruing Ethical Debt that will bankrupt you later.

The Minimum Viable Ethics (MVE) Toolkit

1. The One-Sentence Credo

Write it down in your founding documents. It should be simple and restrictive.

- *Example:* "We use AI to surface talent, never to automate rejection."
- *Example:* "We build tools to help small businesses, not to trap them in debt."
- *Why:* This becomes your North Star when the first gray area revenue opportunity appears.

2. The Single Red Line

Pick one clearly costly thing you will refuse to do, even if it is legal and profitable.

- *Examples:* No scraping non-consensual data. No selling data to third parties. No engagement loops targeting minors.
- *Why:* This is your first costly signal. It proves to early hires and investors that your values are real.

3. The Human Brake

Identify the one decision your product makes that could seriously hurt someone (e.g., denying a loan, flagging fraud).

- **The Rule:** For that one decision, build a "Contact Us" button that goes to a founder's inbox.
- *Why:* You don't need a complex appeals system yet. You just need to ensure that no one is crushed by your MVP without you knowing about it.

4. The No Ghosting Rule

Commit to never leaving a high-stakes user in silence. Even if the answer is "No," deliver it with a reason.

- *Why:* Silence is the seed of resentment. Responsiveness is the seed of loyalty.

The Founder's Question

When you are out of money, out of time, and staring at a term sheet or a contract that makes your stomach turn, ask yourself:

> *"If this becomes a case study in five years, what will it say we were willing to fail as?"*

Your code can pivot in a weekend. Your character becomes your permanent brand. Build the moat early.

APPENDIX F: THE AI CONSTITUTION (OVERVIEW AND DOWNLOAD)

A Governance Starter Kit for Leadership

Use this template to move from good intentions to structural constraints. Adapt the language to your industry, but keep the components.

1. The Credo: Our Sovereign Principle

This paragraph is your North Star. It must be short, memorable, and costly if violated.

"We believe our first responsibility is to the people affected by our systems, to our customers and employees, and those indirectly impacted. We will not use AI to exploit, deceive, or discriminate, even when doing so is legal and profitable. We prioritize the dignity of the human over the efficiency of the machine."

2. Red Lines: What We Will Not Do

Define 3–5 bright-line constraints. If a project crosses these lines, the answer is "No," regardless of revenue.

- **No Black Boxes in Dignity Zones:** We will not deploy unexplainable models in domains that materially affect a person's life chances (e.g., lending, hiring, health, housing). Any such system must support user-facing explanations and human appeals.
- **No Non-Consensual Data Exploitation:** We will not purchase, scrape, or infer sensitive personal data from sources where users did not provide informed, revocable consent.
- **No Toxic Engagement Loops:** We will not build systems whose

core business model depends on addiction, polarization, or the psychological manipulation of vulnerable users.

- **No Ghost Revenue:** We will not knowingly retain revenue generated by technical errors or dark patterns. When discovered, such revenue will be refunded, even at the cost of missing targets.
- **No Secrecy on Structural Failures:** We will not hide material incidents (breaches, systemic bias, safety failures) from affected stakeholders. We will disclose, remediate, and conduct a Blameless Postmortem.

3. The Pre-Decision Audit: Enforcement

Before any major launch, contract, or strategic pivot, the leadership team must answer these three questions in writing:

- **The NYT Test:** If every internal email and rationale behind this decision appeared on the front page of the *New York Times* tomorrow, could we defend it without euphemism?
- **The Universalizer Test:** If every company in our sector made this same decision, would the ecosystem be stronger or weaker?
- **The Sovereign Test:** Does this decision violate our Credo or any published Red Line? (If yes, the default answer is no.)

4. Ownership

- **Owner:** Executive Sponsor (e.g., Chief Risk Officer or equivalent).
- **Review:** Annually with the Board of Directors.
- **Exceptions:** Any exception to a Red Line requires explicit board approval and a documented rationale.

APPENDIX G: THE GOLDEN ALGORITHM SCORECARD

A Self-Assessment of Maturity

Use this Scorecard to baseline your organization. This assessment is divided into two parts: the **Operational Score** (how well you execute) and the **Integrity Foundation** (whether your trust system is credible).

Part 1: The Operational Score (The 6 Pillars)

Evaluate your current AI operations against the six pillars of the Golden Algorithm. Be honest. If a practice is ad-hoc or depends on a single hero individual, it is a 1, not a 2.

Use the rubric below to determine your rating for each pillar from 0 (nonexistent) to 3 (embedded discipline) and record your answers in the **Master Scorecard** at the end of this appendix.

Table 7. The Golden Algorithm Assessment

Pillar	0 (Danger)	1 (Weakness)	2 (Strength)	3 (Moat)	Your Score
G – Guard Human Dignity	"Computer Says No." Users have no recourse.	Manual escalation depends on knowing a specific employee.	Defined appeals process with SLAs exists.	Appeals data drives product improvements.	___
O – Operate Transparently	Black Box. No one can explain decisions.	Technical docs exist but are hidden/outdated.	Model Cards are used in internal reviews.	New systems cannot launch without user-facing explanations.	___
L – Limit Harm	Reactive. We fix it after it breaks.	Manual checks exist but are inconsistent.	Defined thresholds trigger automated review.	Automated Circuit Breakers stop drift; human Andon Cords catch context.	___
D – Design with Empathy	We design for the Default User only.	Ad hoc testing with diverse users.	Vulnerable users are part of core design/testing.	Insights from Stress Cases drive innovation strategy.	___
E – Ensure Accountability	Blame is diffuse. "The model did it."	Informal ownership; finger-pointing happens.	Every system has a named Owner of Record.	Blameless Postmortems are the cultural norm.	___
N – Nurture the Common Good	We only measure direct revenue.	Occasional CSR projects separate from product.	Stakeholder Impact Assessments for major launches.	Externality risk is treated as solvency risk.	___
TOTAL OPERATIONAL SCORE				(Max 18)	___

Part 2: The Integrity Foundation

Integrity is not a gradient—it is the bedrock. This section functions as a binary gateway. If you do not have these foundational commitments in place, high scores in Part 1 are meaningless—you are building a castle on sand.

Answer Yes or No for each test.

Table 8. The Integrity Foundation Assessment

Test	Question	Result (Y/N)
The Sovereign Test	Do we have a written Credo that we have used to veto a profitable opportunity in the last 12 months?	___
The Red Line Test	Do we have explicit, published boundaries (e.g., No non-consensual scraping) that we never cross?	___
The Tylenol Test	When we find a hidden error that benefits us (like Ghost Revenue), do we self-correct before we are caught?	___

The Golden Algorithm Master Scorecard

Combine your results from Part 1 and Part 2 below to determine your organization's maturity level.

Table 9. The Golden Algorithm Maturity Scorecard

Metric	Your Result
Integrity Foundation (Did you answer YES to all 3 in Part 2?)	**PASS / FAIL**
GOLDEN Pillars Score (Total sum of pillars G, O, L, D, E, N)	___ **/ 18**

Interpreting Your Score

The Sandcastle (Integrity Foundation = FAIL)

- *Analysis:* Your score in Part 1 is irrelevant. Without the Integrity Foundation, your processes are performance art. You will likely abandon them when the market turns.

- *Action:* Stop optimizing. Start by drafting your **AI Constitution** (Appendix F).

Fragile (Score 0–8)

- *Analysis:* You have good intentions but weak controls. A serious incident (bias, leak, backlash) will likely overwhelm you.
- *Action:* Pick one pillar (like **L - Limit Harm**) and get it to a Level 2 immediately.

Emerging (Score 9–13)

- *Analysis:* You have tools, but they are reactive. You catch problems, but often too late.
- *Action:* Focus on automating safeguards and assigning clear ownership (**E - Ensure Accountability**) to move faster.

The Fortress (Score 14–18)

- *Analysis:* You have a durable Ethical Moat. Your structure is now a competitive advantage that lowers your Cost of Sales and Talent Acquisition.
- *Action:* Focus on **N - Nurture the Common Good**. Use your surplus trust to lead the market on standards.

Acknowledgements

First and foremost, thank you to my wife, **Katie**, and our sons, **John Oliver ("Jack")** and **James Alexander**. Writing a book is never a solitary act; it is a tax on the time and energy of an entire family. Thank you for your patience, your grace, and for supporting the effort and deep focus required for writing, editing, and thinking during this process. You are the reason I care about the future we are building.

I am deeply grateful to my academic home, **Old Dominion University**. To the leadership of the **Strome College of Business**—specifically Dean Erika Marsillac, Associate Dean Yuping Liu-Thompkins, and Chuanyi Tang, the Chair of the Marketing Department—thank you for your forward-thinking support. You didn't just allow me to explore the intersection of AI and business; you actively championed the expansion of our curriculum to meet this moment. To my amazing colleagues in the Marketing Department and across the College, thank you for the support, the debates, and the shared mission. I also want to recognize the university leadership for seeing the wave before it broke. I also want to recognize Provost Brian Alexander, Dean Holly Handley, and ODU Global for creating the AI Teaching Fellowship, another catalyst in spurring me forward in this work.

Finally, in the spirit of the **Operate Transparently (O)** pillar that I advocate for in this book, I must acknowledge the role of artificial intelligence in the creation of this manuscript. AI can be an effective productivity multiplier, and I used it as such. While the Golden Algorithm, the strategic framework, the ideas, and the ethical convictions in this book are my own, I utilized generative AI as a valuable partner in the execution, as a research assistant and collaborator.

As a thinker with many ideas (and perhaps a tendency toward wordiness), I relied on this assistant to help me organize complex thoughts into a cohesive narrative. It served as a tireless editor, a debate partner, and a creative partner in visualizing concepts throughout the text. It allowed me to focus on what to say, while it helped me refine how to say it. This book is a testament to the idea that when human intent directs machine capability, the result is not replacement, but elevation.

About the Author

Dr. Ryan Baltrip works at the intersection of business strategy, human-centered technology, and education. He is a marketing faculty member and Executive Director of the Loyalty Science Lab at Old Dominion University, where his work focuses on how trust is built, measured, and sustained in modern markets.

Ryan also founded Navigate AI, a platform that helps institutions and leaders develop practical AI fluency, moving beyond tool adoption to governance, decision discipline, and strategy. His writing blends real-world leadership constraints with a simple conviction: in the age of AI, the organizations that win long-term will be the ones that can scale capability without scaling harm.

You can find related resources, frameworks, and training at:

navigateai.org

Endnotes

Preface

1 Matthew 7:12 and/or Luke 6:31, NIV.

2 Ecclesiastes 1:9, NIV.

Introduction

3 U.S. Securities and Exchange Commission and Commodity Futures Trading Commission, *Findings Regarding the Market Events of May 6, 2010: Report of the Staffs to the Joint Advisory Committee on Emerging Regulatory Issues*, September 30, 2010.

4 *Ibid.* The joint staff report found that a large automated sell program in E-mini S&P 500 futures, interacting with thinning liquidity and automated responses, helped amplify the event.

5 Paul Virilio, *Speed and Politics*, trans. Mark Polizzotti (New York: Semiotext(e), 2006). And then Langdon Winner, "Do Artifacts Have Politics?" *Daedalus* 109, no. 1 (1980): 121–136. Winner argues that technical systems are not neutral tools but political structures that enforce specific values.

6 Henry Blodget, "Mark Zuckerberg: On Innovation," *Business Insider*, October 1, 2009, https://www.businessinsider.com/mark-zuckerberg-innovation-2009-10 (accessed January 2, 2026). This motto, originally internal to Facebook, came to define an entire era of software development where speed was valued over stability or safety.

7 Joseph E. Stiglitz, "Information and the Change in the Paradigm in Economics," *American Economic Review* 92, no.3 (June 2002).

8 Jeremy Rifkin, *The Zero Marginal Cost Society* (New York: Palgrave Macmillan, 2014).

9 A variety of ideas influenced the thinking here: George A. Akerlof, "The Market for 'Lemons': Quality Uncertainty and the Market Mechanism," *The Quarterly Journal of Economics* 84, no. 3 (1970); Michael Spence, "Job Market Signaling," *The Quarterly Journal of Economics* 87, no. 3 (1973); and Henry Dunning Macleod, *The Elements of Political Economy* (London: Longman, 1858), 476. Applied to information markets, this suggests that cheap, low-quality content (generated AI) will drive out expensive, high-quality verification unless a signaling mechanism exists.

10 Rachel Botsman, *Who Can You Trust? How Technology Brought Us Together and Why It Could Drive Us Apart* (New York: PublicAffairs, 2017). Botsman argues that trust has shifted from "institutional" (banks, governments) to "distributed" (peers, platforms), creating a volatility that defines the Trust Recession; Edelman, *2024 Edelman Trust Barometer: Global Report* (Chicago: Edelman Holdings, 2024).

11 Daniel Kahneman, *Thinking, Fast and Slow* (New York: Farrar, Straus and Giroux, 2011).

12 Emily M. Bender et al., "On the Dangers of Stochastic Parrots: Can Language Models Be Too Big?," *FAccT '21* (ACM, 2021), 610–623. The authors argue that Large Language Models do not understand meaning but merely mimic the statistical likelihood of word sequences, creating a risk of plausible-sounding falsehoods.

13 *Ibid.*

14 Shoshana Zuboff, *The Age of Surveillance Capitalism* (New York: PublicAffairs, 2019).

Chapter 1

15 Nick Bostrom, *Superintelligence: Paths, Dangers, Strategies* (Oxford: Oxford University Press, 2014), 123. The 'Paperclip Maximizer' is a thought experiment illustrating how an AI system with a harmless goal (making paperclips) can destroy the world if it lacks countervailing values.

16 Richard Rothstein, *The Color of Law: A Forgotten History of How Our Government Segregated America* (New York: Liveright Publishing Corporation, 2017).

17 Jongwoon Choi, Gary W. Hecht, and William B. Tayler, "Strategy Selection, Surrogation, and Strategic Performance Measurement Systems," *The Accounting Review* 87, no. 4, pp. 1135-1163 (2012). Surrogation occurs when managers psychologically replace the abstract strategy (e.g., 'customer satisfaction') with the metric used to measure it (e.g., 'NPS score').

18 Charles A. E. Goodhart, "Problems of Monetary Management: The U.K. Experience," in *Monetary Economics*, ed. A. S. Courakis (London: Macmillan Education UK, 1975), 91–121. Often summarized as: 'When a measure becomes a target, it ceases to be a good measure.

19 The term "Cobra Effect" stems from an anecdote about British rule in India, though the most well-documented historical example is the French colonial "Rat Hunt" in Hanoi. For a detailed history of the perverse incentives in Hanoi, see Michael G. Vann, *The Great Hanoi Rat Hunt: Empire, Disease, and Modernity in French Colonial Vietnam* (Oxford University Press, 2018).

20 Scott Reckard, "Wells Fargo's Pressure-Cooker Sales Culture Comes at a Cost," *The Los Angeles Times*, December 21, 2013; Consumer Financial Protection Bureau (CFPB), "Consumer Financial Protection Bureau Fines Wells Fargo $100 Million for Widespread Illegal Practice of Secretly Opening Unauthorized Accounts," news release, September 8, 2016.

21 Independent Directors of the Board of Wells Fargo & Company, *Sales Practices Investigation Report* (April 10, 2017), 12–15. The report details how the "Eight is Great" cross-sell metric became the primary driver of unethical behavior.

22 Michael Corkery and Stacy Cowley, "Wells Fargo's Pressure-Cooker Sales Culture Comes at a Cost," *The New York Times*, September 8, 2016; Consumer Financial Protection Bureau (CFPB), "Consumer Financial Protection Bureau Fines Wells Fargo $100 Million for Widespread Illegal Practice of Secretly Opening Unauthorized Accounts," news release, September 8, 2016.

23 John W. Meyer and Brian Rowan, "Institutionalized Organizations: Formal Structure as Myth and Ceremony," *American Journal of Sociology* 83, no. 2 (1977): 340-363. Strategic Decoupling refers to the gap between an organization's formal policies (what they say) and their actual daily practices (what they do).

24 Frederick Winslow Taylor, *The Principles of Scientific Management* (New York: Harper & Brothers, 1911).

25 Annette Bernhardt et al., "Data and Algorithms at Work: The Case for Worker Technology Rights," *ILR Review* 76, no. 1 (2023). The authors detail how algorithmic management tools enable a level of surveillance and control that exceeds the physical limitations of human managers.

26 Strategic Organizing Center (SOC), "The Injury Machine: How Amazon's Production System Hurts Workers" (report), April 20, 2022.

27 Gene Kim, Kevin Behr, and George Spafford, *The Phoenix Project: A Novel about IT, DevOps, and Helping Your Business Win* (Portland, OR: IT Revolution Press, 2013). The authors illustrate the "Wait Time" curve, showing how as resource utilization approaches 100%, wait times (latency) approach infinity.

28 Taiichi Ohno, *Toyota Production System: Beyond Large-Scale Production* (Cambridge, MA: Productivity Press, 1988).

29 Edward Tenner, *Why Things Bite Back: Technology and the Revenge of Unintended Consequences* (New York: Knopf, 1996).

30 National Institute of Standards and Technology (NIST), *Artificial Intelligence Risk Management Framework (AI RMF 1.0)* (U.S. Dept. of Commerce, 2023); International Organization for Standardization (ISO) and International Electrotechnical Commission (IEC), *ISO/IEC 42001:2023 — Information technology — Artificial intelligence — Management system* (2023).

Chapter 2

31 George A. Akerlof, "The Market for 'Lemons': Quality Uncertainty and the Market Mechanism," *The*

Quarterly Journal of Economics 84, no. 3 (1970). Akerlof demonstrated that when buyers cannot distinguish quality, they assume the worst (all cars are 'lemons'), driving high-quality sellers out of the market. This is the central threat to the AI economy

32 Joseph E. Stiglitz, "Information and the Change in the Paradigm in Economics," *American Economic Review* 92, no. 3 (2002): 460–501. Stiglitz's Nobel work on information asymmetry explains how market failures occur when one party (the AI builder) knows more about the risk than the other (the user).

33 Henry Dunning Macleod, *The Elements of Political Economy* (London: Longman, 1858), 476. Macleod popularized "Gresham's Law" (bad money drives out good). In AI, this implies that cheap, generative noise will drive out high-quality, verified signals unless a mechanism for trust exists.

34 Rachel Botsman, *Who Can You Trust? How Technology Brought Us Together and Why It Could Drive Us Apart* (New York: PublicAffairs, 2017). Botsman argues that trust has shifted from "institutional" (banks, governments) to "distributed" (peers, platforms), creating a volatility that defines the Trust Recession.

35 Michael Spence, "Job Market Signaling," *The Quarterly Journal of Economics* 87, no. 3 (1973). Spence argued that for a signal to be effective in resolving uncertainty, it must be costly to the signaler (e.g., a degree, a warranty, or a public audit).

36 Jeremy Rifkin, *The Zero Marginal Cost Society: The Internet of Things, the Collaborative Commons, and the Eclipse of Capitalism* (New York: Palgrave Macmillan, 2014). Rifkin predicted that technology would drive the cost of producing information to near zero, making "cheap talk" infinite and value harder to capture.

37 Information Commissioner's Office (UK), "Data as a commodity," May 26, 2017, https://ico.org.uk/for-the-public/ico-40/data-as-a-commodity/ (accessed November 12, 2025). This ICO note documents the attribution to Clive Humby in a speech at the 2006 Association of National Advertisers Conference.

38 See *Meinhard v. Salmon*, 249 N.Y. 458, 464 (1928). Justice Cardozo's opinion defined the fiduciary standard as "the punctilio of an honor the most sensitive," establishing that a fiduciary must subordinate their own interest to that of the beneficiary.

39 Apple Inc., "iOS 14.5 Offers Unlock iPhone with Apple Watch, More Diverse Siri Voice Options, and New Privacy Controls," press release, April 26, 2021, https://www.apple.com/newsroom/2021/04/ios-14-5-offers-unlock-iphone-with-apple-watch-diverse-siri-voices-and-more/ (accessed on November 2, 2025).

40 Meta Platforms, Inc., "Q4 2021 Earnings Call Transcript" (Feb. 2, 2022), https://s21.q4cdn.com/399680738/files/doc_financials/2021/q4/Meta-Q4-2021-Earnings-Call-Transcript.pdf (accessed November 15, 2025). See also Meta Platforms, Inc., "Q1 2022 Earnings Call Transcript" (Apr. 27, 2022), https://s21.q4cdn.com/399680738/files/doc_financials/2022/q1/Meta-Q1-2022-Earnings-Call-Transcript.pdf (accessed November 15, 2025). Meta discusses an estimated ~$10B 2022 "headwind" tied to Apple iOS/App Tracking Transparency changes.

41 Warren Buffett, "Chairman's Letter," in *Berkshire Hathaway Annual Report 1995* (February 29, 1996). Buffett defines an 'economic moat' as a structural competitive advantage that protects a business from competitors, often citing brand trust and switching costs.

42 For a theoretical basis on transparency as a competitive advantage, see: Don Tapscott and David Ticoll, *The Naked Corporation: How the Age of Transparency Will Revolutionize Business* (New York: Free Press, 2003).

43 Angela Glover Blackwell, "The Curb-Cut Effect," *Stanford Social Innovation Review* 15, no. 1 (Winter 2017): 28–33. The article details how accommodations for the disabled (curb cuts) ended up benefiting the entire population (strollers, travelers), a metaphor for designing inclusive AI.

Chapter 3

44 Immanuel Kant, *Groundwork of the Metaphysics of Morals* (Cambridge University Press, 1998), 4:429. The second formulation of the Categorical Imperative states: "Act in such a way that you treat humanity, whether in your own person or in the person of any other, never merely as a means to an end, but always at the same time as an end."

45 For the psychological basis of objectification, see: Barbara L. Fredrickson and Tomi-Ann Roberts, "Objectification Theory: Toward Understanding Women's Lived Experiences and Mental Health Risks," *Psychology of Women Quarterly* 21, no. 2 (1997): 173–206.

46 Colin Lecher, "How Amazon automatically tracks and fires warehouse workers," *The Verge*, April 25, 2019.

47 This mirrors the "Inadvertent Algorithmic Cruelty" concept coined by Eric Meyer. See: Eric Meyer, "Inadvertent Algorithmic Cruelty," *Eric's Archived Thoughts* (blog), December 24, 2014. (See note in Intro).

48 *Little Britain*, Season 2, Episode 1, "Computer Says No," aired October 19, 2004, BBC Three. The sketch satirizes the helplessness of frontline workers who cannot override a system's output, a dynamic now common in AI-mediated service.

49 *Magna Carta Libertatum*, Clause 39 (1215).

50 This principle is codified in modern law as Article 22 of the GDPR: "The data subject shall have the right not to be subject to a decision based solely on automated processing." *Regulation (EU) 2016/679* (General Data Protection Regulation), Art. 22(1).

51 *Regulation (EU) 2016/679* (General Data Protection Regulation), Article 22 ("Automated individual decision-making, including profiling") and Recital 71, Official Journal of the European Union, April 27, 2016, https://eur-lex.europa.eu/eli/reg/2016/679/oj (accessed November 17, 2025). For the contested scope of a "right to explanation," see Sandra Wachter, Brent Mittelstadt, and Luciano Floridi, "Why a Right to Explanation of Automated Decision-Making Does Not Exist in the General Data Protection Regulation," International Data Privacy Law 7, no. 2 (2017): 76–99.

52 Anna Brose (@alcesanna), "I contacted @Chewy last week ..." post on X (formerly Twitter), June 15, 2022, https://x.com/alcesanna/status/1536930380082728961 (accessed November 18, 2025). The tweet received over 730,000 likes and sparked thousands of replies from other customers sharing similar stories of Chewy's "grief logistics." See also Tim Bella, "Her dog died suddenly. Then a Chewy delivery brought a surprise," Washington Post, June 18, 2022, https://www.washingtonpost.com/lifestyle/2022/06/18/dog-gus-chewy-brose-twitter/ (accessed November 18, 2025).

53 This echoes Max Weber's concept of the "Iron Cage" of rationalized bureaucracy. See: Max Weber, *The Protestant Ethic and the Spirit of Capitalism*, trans. Talcott Parsons (New York: Scribner, 1958). Weber feared that the rationalization of society would trap humanity in an 'iron cage' of bureaucratic control, stripping away individual agency.

54 For standard definitions of SLAs in computing, see: Peter Mell and Timothy Grance, "The NIST Definition of Cloud Computing," *NIST Special Publication* 800-145 (2011).

Chapter 4

55 Tom Fawcett, "An Introduction to ROC Analysis," *Pattern Recognition Letters* 27, no. 8 (2006): 861–874. AUC (Area Under the Curve) is a standard metric for evaluating the performance of a classification model, but it often masks performance failures for specific subgroups.

56 Marco Tulio Ribeiro, Sameer Singh, and Carlos Guestrin, "'Why Should I Trust You?': Explaining the Predictions of Any Classifier," in *Proceedings of the 22nd ACM SIGKDD International Conference on Knowledge Discovery and Data Mining* (New York: ACM, 2016), 1135–1144, https://doi.org/10.1145/2939672.2939778 (accessed October 3, 2025).

57 Frank Pasquale, *The Black Box Society: The Secret Algorithms That Control Money and Information* (Cambridge, MA: Harvard University Press, 2015).

58 *The Financial Crisis Inquiry Report: Final Report of the National Commission on the Causes of the Financial and Economic Crisis in the United States* (Washington, DC: Government Printing Office, 2011).

59 Luca Pacioli, *Summa de Arithmetica, Geometria, Proportioni et Proportionalita* (Venice: Paganino of Paganini, 1494). Pacioli's codification of double-entry bookkeeping created a system where errors became visible, allowing strangers to trust a ledger they did not personally write.

60 Margaret Mitchell et al., "Model Cards for Model Reporting," in *Proceedings of the Conference on Fairness, Accountability, and Transparency* (ACM, 2019), 220–229. A Model Card is a standardized document that explains a model's intended use, limitations, and performance across different demographic groups, effectively creating a 'nutrition label' for AI.

61 Scott M. Lundberg and Su-In Lee, "A Unified Approach to Interpreting Model Predictions," in *Advances in Neural Information Processing Systems 30* (NeurIPS 2017), 4765–4774, https://doi.org/10.48550/arXiv.1705.07874 (accessed October 4, 2025).

62 See Sandra Wachter, Brent Mittelstadt, and Chris Russell, "Counterfactual Explanations without Opening the Black Box: Automated Decisions and the GDPR," *Harvard Journal of Law & Technology* 31, no. 2 (2018): 841–887.

63 Emmanuel Martinez and Lauren Kirchner, "The Secret Bias Hidden in Mortgage-Approval Algorithms," *The Markup* and *Associated Press*, August 25, 2021.

64 Nassim Nicholas Taleb, *Antifragile: Things That Gain from Disorder* (New York: Random House, 2012). Taleb distinguishes "antifragility" (getting stronger from shock) from "resilience" (merely withstanding shock).

65 Amy Edmondson, *The Fearless Organization: Creating Psychological Safety in the Workplace* (Hoboken, NJ: Wiley, 2018).

66 James C. Scott, *Seeing Like a State: How Certain Schemes to Improve the Human Condition Have Failed* (New Haven, CT: Yale University Press, 1998). Scott argues that large organizations (states) force complex, messy reality into simplified, 'legible' formats to manage it, often destroying local knowledge in the process.

Chapter 5

67 Defense Science Board Task Force, *The Role and Status of DoD Red Teaming Activities* (Washington, DC: Office of the Under Secretary of Defense, 2003). "Red Teaming" originated in military wargaming as a method to challenge plans and assumptions from an adversarial perspective.

68 Diane Vaughan, *The Challenger Launch Decision: Risky Technology, Culture, and Deviance at NASA* (Chicago: University of Chicago Press, 1996). Vaughan defines "Normalization of Deviance" as the process where a group gradually accepts technical anomalies as normal because no immediate catastrophe occurs. She argues that the disaster was not caused by rule-breaking, but by rule-following within a flawed culture.

69 Charles Perrow, *Normal Accidents: Living with High-Risk Technologies* (New York: Basic Books, 1984). Perrow argues that in "tightly coupled" complex systems, accidents are not anomalies but inevitable ("normal") outcomes of the system's design.

70 Betsy Beyer et al., eds., *Site Reliability Engineering: How Google Runs Production Systems* (Sebastopol, CA: O'Reilly Media, 2016). In software engineering, blast radius refers to the scope of impact of a specific failure.

71 Tali Sharot, The Optimism Bias: A Tour of the Irrationally Positive Brain (New York: Vintage, 2012). Sharot details the biological basis for the "optimism bias," which often leads project teams to underestimate the likelihood of negative events.

72 Gary Klein, "Performing a Project Premortem," *Harvard Business Review* 85, no. 9 (September 2007): 18–19. Unlike a postmortem, which asks "what went wrong," a premortem operates on the hypothetical certainty that the project *has already failed* and asks the team to generate the cause.

73 Taiichi Ohno, *Toyota Production System: Beyond Large-Scale Production* (Cambridge, MA: Productivity Press, 1988). Ohno describes the "stop-the-line" authority as essential to *jidoka* (automation with a human touch).

74 NASA, *System Safety Handbook, Volume 1: System Safety Framework and Concepts for Implementation* (Washington, DC: NASA, 2011). NASA defines "fail-safe" as a design feature that ensures a system defaults to a safe condition in the event of a failure (e.g., a train brake that engages when pressure is lost).

75 U.S. Securities and Exchange Commission, "Investor Bulletin: New Measures to Address Market Volatility," April 9, 2013. https://www.sec.gov/investor/alerts/circuitbreakersbulletin.htm (accessed December 13, 2025). Following the 1987 crash, markets instituted "circuit breakers" to mandate cooling-off periods during extreme volatility.

76 Karl E. Weick and Kathleen M. Sutcliffe, *Managing the Unexpected: Assuring High Performance in an Age of Complexity* (San Francisco: Jossey-Bass, 2001). Weick and Sutcliffe studied "High Reliability Organizations" (HROs) like aircraft carriers and nuclear plants to understand how they manage to operate nearly error-free in unforgiving environments.

77 Sidney Dekker, *Just Culture: Balancing Safety and Accountability* (Aldershot, UK: Ashgate, 2007). Dekker argues that punishing employees for honest mistakes creates a culture of secrecy that ultimately increases systemic risk. A "Just Culture" draws a sharp line between error (to be learned from) and recklessness (to be punished).

78 This maxim is widely attributed to the U.S. Navy SEALs and U.S. Army Special Forces, emphasizing that deliberate precision prevents the errors that ultimately slow down an operation.

Chapter 6

79 Jack Halberstam, *The Queer Art of Failure* (Durham, NC: Duke University Press, 2011). Halberstam critiques the culture of "toxic positivity" that pathologizes negative emotion and failure, arguing that systems often enforce happiness to mask structural inequality.

80 Joseph Henrich, Steven J. Heine, and Ara Norenzayan, "The Weirdest People in the World?" *Behavioral and Brain Sciences* 33, no. 2–3 (2010): 61–83. The authors define "WEIRD" (Western, Educated, Industrialized, Rich, and Democratic) populations, noting that 96% of psychological samples come from countries with only 12% of the world's population.

81 For the origin of Survivorship Bias, see: Howard Wainer, *Visual Revelations: Graphical Tales of Fate and Deception from Napoleon Bonaparte to Ross Perot* (New York: Copernicus, 1997). Wainer recounts the story of Abraham Wald, who realized that analyzing bullet holes on returning WWII planes was flawed because it ignored the planes that didn't return.

82 Don Norman, *The Design of Everyday Things* (New York: Basic Books, 1988). Norman critiques the tendency of engineers to design for the "ideal" logic of the machine rather than the messy reality of human behavior.

83 Emily M. Bender et al., "On the Dangers of Stochastic Parrots: Can Language Models Be Too Big?" in *Proceedings of the 2021 ACM Conference on Fairness, Accountability, and Transparency* (FAccT '21) (New York: ACM, 2021), 610–623, https://doi.org/10.1145/3442188.3445922 (accessed November 22, 2025). The authors argue that LLMs, by averaging vast amounts of data, tend to reproduce the dominant (hegemonic) view, smoothing over the specific, jagged realities of marginalized groups.

84 For a critique of emotional empathy as a guide for moral action, see: Paul Bloom, *Against Empathy: The Case for Rational Compassion* (New York: Ecco, 2016). Bloom argues that "cognitive empathy" (understanding) is more reliable than "emotional empathy" (feeling).

85 For a history of the Rolling Quads and the independent living movement, see: Joseph P. Shapiro, *No Pity: People with Disabilities Forging a New Civil Rights Movement* (New York: Times Books, 1993). The first official curb cuts were installed in Berkeley in 1972 on Telegraph Avenue.

86 Angela Glover Blackwell, "The Curb-Cut Effect," *Stanford Social Innovation Review* 15, no. 1 (Winter 2017): 28–33. Blackwell argues that laws and programs designed for vulnerable groups often end up benefiting society as a whole.

87 Ronald L. Mace, "Universal Design: Barrier Free Environments for Everyone," *Designers West* 33, no. 1 (1985): 147–152. Mace coined the term "Universal Design" to describe products and environments usable by all people, to the greatest extent possible, without the need for adaptation.

88 WebAIM, "The WebAIM Million: The 2023 Report on the Accessibility of the Top 1,000,000 Home Pages" (2023), https://webaim.org/projects/million/2023/ (accessed November 23, 2025). The report found that 96.3% of the top one million home pages had detectable WCAG 2 failure, demonstrating systemic exclusion.

89 Mark Wilson, "The Untold Story of the Vegetable Peeler That Changed the World," *Fast Company*, September 24, 2018.

90 Microsoft, *Inclusive Design Toolkit* (2016). Microsoft distinguishes "Inclusive Design" (solving for one extends to many) from "Universal Design" (one solution for everyone).

91 Defense Science Board Task Force, *The Role and Status of DoD Red Teaming Activities* (Washington, DC: Office of the Under Secretary of Defense, 2003).

Chapter 7

92 John M. Darley and Bibb Latané, "Bystander Intervention in Emergencies: Diffusion of Responsibility," *Journal of Personality and Social Psychology* 8, no. 4 (1968): 377–383.

93 James Reason, *Human Error* (Cambridge: Cambridge University Press, 1990). Reason's "Swiss Cheese Model" illustrates how accidents occur only when holes in multiple layers of defense align, validating the idea that "the system," not only the person, is often to blame.

94 Michael C. Jensen and William H. Meckling, "Theory of the Firm: Managerial Behavior, Agency Costs and Ownership Structure," *Journal of Financial Economics* 3, no. 4 (1976): 305–360. The "Principal-Agent Problem" explains why employees (agents) optimized for speed often take risks that harm the long-term health of the firm (principal).

95 Madeline Clare Elish, "Moral Crumple Zones: Cautionary Tales in Human-Robot Interaction," *Engaging Science, Technology, and Society* 5 (2019): 40–60, https://doi.org/10.17351/ests2019.260 (accessed January 2, 2026).

96 For a precedent on "Algorithmic Disgorgement," see: *In the Matter of Kurbo Inc. and WW International, Inc.*, FTC File No. 1923091 (2022). The settlement required the companies to delete not just the data collected illegally, but the algorithms trained on that data, a powerful incentive for accountability.

97 Nassim Nicholas Taleb, *Skin in the Game: Hidden Asymmetries in Daily Life* (New York: Random House, 2018). Taleb argues that systems become fragile when decision-makers are insulated from the consequences of their decisions.

98 Project Management Institute, *A Guide to the Project Management Body of Knowledge (PMBOK Guide)*, 7th ed. (Newtown Square, PA: PMI, 2021).

99 Sidney Dekker, *Just Culture: Balancing Safety and Accountability* (Aldershot, UK: Ashgate, 2007). Dekker argues that a "Just Culture" is not about letting people off the hook, but about aligning the consequence with the behavior (error vs. recklessness).

100 National Transportation Safety Board, *Major Investigations Manual*, Appendix A: The Party System (Washington, DC: NTSB, 2002). The NTSB uses a "Party System" to include technical experts from manufacturers in the investigation without allowing them to control the findings, ensuring learning takes precedence over liability.

101 Robert B. Cialdini, *Influence: The Psychology of Persuasion* (New York: Harper Business, 2006). Cialdini identifies "Consistency" as a key principle of persuasion: once we take a public stand, we encounter personal and interpersonal pressures to behave consistently with that commitment.

Chapter 8

102 Natasha Dow Schüll, *Addiction by Design: Machine Gambling in Las Vegas* (Princeton, NJ: Princeton University Press, 2012). Schüll's anthropological study details how machine intelligence is engineered to create a "zone" of suspended reality that maximizes "time on device" at the expense of user agency.

103 Langdon Winner, "Do Artifacts Have Politics?" *Daedalus* 109, no. 1 (1980): 121–136.

104 James J. Gibson, *The Ecological Approach to Visual Perception* (Boston: Houghton Mifflin, 1979). Gibson introduced the concept of "affordances," which is the action possibilities that the environment offers the animal. In AI design, the algorithm "affords" certain behaviors (addiction, outrage) while discouraging others.

105 Garrett Hardin, "The Tragedy of the Commons," *Science* 162, no. 3859 (1968): 1243–1248. And Anna C. Merritt, Daniel A. Effron, and Benoit Monin, "Moral Self-Licensing: When Being Good Frees Us to

Be Bad," *Social and Personality Psychology Compass* 4, no. 5 (2010): 344–357. The authors explain how a history of "good deeds" can psychologically license a leader to make an unethical choice later.

106 Arthur C. Pigou, *The Economics of Welfare* (London: Macmillan, 1920). Pigou used the famous example of a factory's smoke causing uncompensated damage to neighboring laundry.

107 For the concept of "Moral Limits of Markets," see: Michael J. Sandel, *What Money Can't Buy: The Moral Limits of Markets* (New York: Farrar, Straus and Giroux, 2012). Sandel argues that market values crowd out non-market norms, eroding the common good.

108 John Doerr, *Measure What Matters: How Google, Bono, and the Gates Foundation Rock the World with OKRs* (New York: Portfolio, 2018). Doerr distinguishes between "mercenaries" (driven by money) and "missionaries" (driven by purpose), arguing that missionaries build more valuable companies.

109 For the foundational application of adverse selection to workforce dynamics, see: Bruce C. Greenwald, "Adverse Selection in the Labour Market," *The Review of Economic Studies* 53, no. 3 (1986): 325–347. Greenwald applies the "Lemons" model to hiring, demonstrating how firms that cannot accurately value or retain high-quality employees eventually end up with a workforce dominated by "lemons" (lower-quality workers) as the top talent exits.

110 For a standard treatment of Discounted Cash Flow (DCF) analysis and valuation, see: Aswath Damodaran, *Investment Valuation: Tools and Techniques for Determining the Value of Any Asset* (Hoboken, NJ: Wiley, 2012).

111 R. Edward Freeman, *Strategic Management: A Stakeholder Approach* (Boston: Pitman, 1984). Freeman's Stakeholder Theory argues that a company must create value for all groups without whose support the organization would cease to exist, not just for the shareholders.

112 Shoshana Zuboff, *The Age of Surveillance Capitalism: The Fight for a Human Future at the New Frontier of Power* (New York: PublicAffairs, 2019). Zuboff describes the "behavioral modification" business model, where human experience is claimed as free raw material for translation into behavioral data.

113 Aristotle, *Nicomachean Ethics*, trans. W.D. Ross (Oxford: Clarendon Press, 1908). Teleology (*telos*) refers to the ultimate purpose or goal of a thing.

114 See *Donoghue v. Stevenson*, [1932] AC 562 (House of Lords). This foundational tort case established the "neighbor principle," which means that you must take reasonable care to avoid acts or omissions which you can reasonably foresee would be likely to injure your neighbor.

115 "The Techlash Against Amazon, Facebook and Google—and What They Can Do," *The Economist*, January 20, 2018.

116 European Parliament, "EU AI Act: First Regulation on Artificial Intelligence," European Parliament News, June 8, 2023, updated February 19, 2025, https://www.europarl.europa.eu/topics/en/article/20230601STO93804/eu-ai-act-first-regulation-on-artificial-intelligence (accessed December 14, 2025); Kathryn E. Spier and Xinyu Hua, "Holding Platforms Liable" (paper presented at the Federal Trade Commission 16th Annual Microeconomics Conference, Washington, D.C., November 2023), https://www.ftc.gov/system/files/ftc_gov/pdf/spierhua.pdf (accessed December 14, 2025); Polona Car and Filippo Cassetti, "Regulating Dark Patterns in the EU: Towards Digital Fairness," European Parliamentary Research Service (EPRS), At a Glance, PE 767.191 (2025), https://www.europarl.europa.eu/RegData/etudes/ATAG/2025/767191/EPRS_ATA(2025)767191_EN.pdf (accessed December 14, 2025); and Fabiana Di Porto and Alexander Egberts, "The Collective Welfare Dimension of Dark Patterns Regulation," *European Law Journal* 29, no. 1–2 (2023): 114–141.

Chapter 9

117 Bernard Williams, "A Critique of Utilitarianism," in *Utilitarianism: For and Against* (Cambridge: Cambridge University Press, 1973). Williams argues that utilitarianism alienates individuals from their own moral integrity by asking them to be mere vessels for the "aggregate good."

118 Dan Ariely, *The (Honest) Truth About Dishonesty: How We Lie to Everyone—Especially Ourselves* (New York: HarperCollins, 2012). Ariely describes the "fudge factor," which is the ability of honest people to cheat just enough to profit, but not enough to damage their own self-image.

119 Mark Dowie, "Pinto Madness," *Mother Jones*, September/October 1977. This investigative report exposed the internal Ford memo that calculated the cost of fixing a fuel tank defect ($11 per car) against the cost of paying off burn victims, concluding it was cheaper to let the cars burn.

120 Ward Cunningham, "The WyCash Portfolio Management System," *ACM Sigplan Oops Messenger* 4, no. 2 (1992): 29-30. See also Martin Fowler, "Technical Debt," May 21, 2019, https://martinfowler.com/bliki/TechnicalDebt.html (accessed December 17, 2025), and Steve McConnell, Managing Technical Debt (Construx Software Best Practices White Paper, Version 1, June 2008), https://www.construx.com/uploadedfiles/resources/whitepapers/Managing%20Technical%20Debt.pdf (accessed December 17, 2025). Cunningham coined the metaphor "Technical Debt" to explain how rushing code out the door creates an invisible cost that must be paid down later (via rework), or else it accrues "interest" (complexity and bugs). Here, we are using technical debt as an analogy for "Ethical Debt." Just as technical debt accrues interest in the form of bugs and fragility, Ethical Debt accrues interest in the form of regulatory risk and talent flight.

121 Benjamin Klein and Keith B. Leffler, "The Role of Market Forces in Assuring Contractual Performance," *Journal of Political Economy* 89, no. 4 (1981): 615–641. This economic paper demonstrates that high-quality firms must charge a premium (reputational capital) to make "cheating" economically irrational.

122 U.S. Environmental Protection Agency, "Notice of Violation of the Clean Air Act" (Volkswagen AG), September 18, 2015. The "Dieselgate" scandal is the definitive modern example of embedding deception into code to bypass regulation.

123 Anna C. Merritt, Daniel A. Effron, and Benoit Monin, "Moral Self-Licensing: When Being Good Frees Us to Be Bad," *Social and Personality Psychology Compass* 4, no. 5 (2010): 344–357. The authors explain how a history of "good deeds" can psychologically license a leader to make an unethical choice later.

124 Homer, *The Odyssey*, trans. Robert Fagles (New York: Penguin Books, 1996), Book 12. In modern decision theory, this concept helps explain pre-commitment strategies.

125 Judith Rehak, "Tylenol Made a Hero of Johnson & Johnson: The Recall That Started Them All," *The New York Times*, March 23, 2002. Burke's decision to recall 31 million bottles cost $100 million but saved the brand.

126 Amotz Zahavi, "Mate Selection—A Selection for a Handicap," *Journal of Theoretical Biology* 53, no. 1 (1975): 205–214. Zahavi argues that reliable signals must be costly to the signaler (like a peacock's tail) to prove they are not faking.

Chapter 10

127 Frederick Winslow Taylor, *The Principles of Scientific Management* (New York: Harper & Brothers, 1911).

128 Ward Cunningham, "The WyCash Portfolio Management System," *ACM Sigplan Oops Messenger* 4, no. 2 (1992): 29-30. See also Martin Fowler, "Technical Debt," May 21, 2019, https://martinfowler.com/bliki/TechnicalDebt.html (accessed December 17, 2025), and Steve McConnell, Managing Technical Debt (Construx Software Best Practices White Paper, Version 1, June 2008), https://www.construx.com/uploadedfiles/resources/whitepapers/Managing%20Technical%20Debt.pdf (accessed December 17, 2025). As mentioned in Chapter 9 note, Ethical Debt accrues interest in the form of regulatory risk and talent flight.

129 Edelman, *2024 Edelman Trust Barometer: Global Report* (Chicago: Daniel J. Edelman Holdings, Inc., 2024).

130 Nassim Nicholas Taleb, *Antifragile: Things That Gain from Disorder* (New York: Random House, 2012). Taleb defines antifragility as the property of systems that benefit from shocks, volatility, and stress, as opposed to fragile systems that break.

131 Dominic Barton and Mark Wiseman, "Focusing Capital on the Long Term," *Harvard Business Review* 92, no. 1 (2014): 44–51. The authors argue that "patient capital" seeks companies with sustainable governance structures, as they outperform quarterly-focused firms over time.

132 Melvin E. Conway, "How Do Committees Invent?" *Datamation* 14, no. 4 (April 1968): 28–31. Con-

way's Law implies that if an organization has a fragmented, siloed culture, its software (and its safety architecture) will be equally fragmented.

133 National Institute of Standards and Technology (NIST), *Artificial Intelligence Risk Management Framework (AI RMF 1.0)* (Washington, DC: U.S. Department of Commerce, 2023). The AI RMF has become the de facto standard for government and enterprise procurement of safe AI systems.

134 Amotz Zahavi, "Mate Selection—A Selection for a Handicap," *Journal of Theoretical Biology* 53, no. 1 (1975): 205–214. As mentioned in Chapter 9 note, Zahavi introduced the "Handicap Principle," arguing that reliable signals must be costly to the signaler to prove they are not faking.

135 **Epilogue**

136 Tensie Whelan, Ulrich Atz, Tracy Van Holt, and Casey Clark, "ESG and Financial Performance: Uncovering the Relationship by Aggregating Evidence from 1,000 Plus Studies Published between 2015-2020," *NYU Stern Center for Sustainable Business* (2021). The meta-analysis found a positive correlation between robust ESG (Environmental, Social, Governance) practices and financial performance in 58% of studies, validating the "integrity premium."

137 Financial Reporting Council, *The UK Stewardship Code 2020* (London: FRC, 2020). The code defines stewardship as "the responsible allocation, management and oversight of capital to create long-term value for clients and beneficiaries leading to sustainable benefits for the economy, the environment and society."

138 Milton Friedman, "The Social Responsibility of Business Is to Increase Its Profits," *The New York Times Magazine*, September 13, 1970. Friedman famously argued that corporate executives who spend money on social causes are effectively imposing a tax on shareholders.

www.ingramcontent.com/pod-product-compliance
Ingram Content Group UK Ltd.
Pitfield, Milton Keynes, MK11 3LW, UK
UKHW020425250726
13967UKWH00007B/2821

9 780998 033341